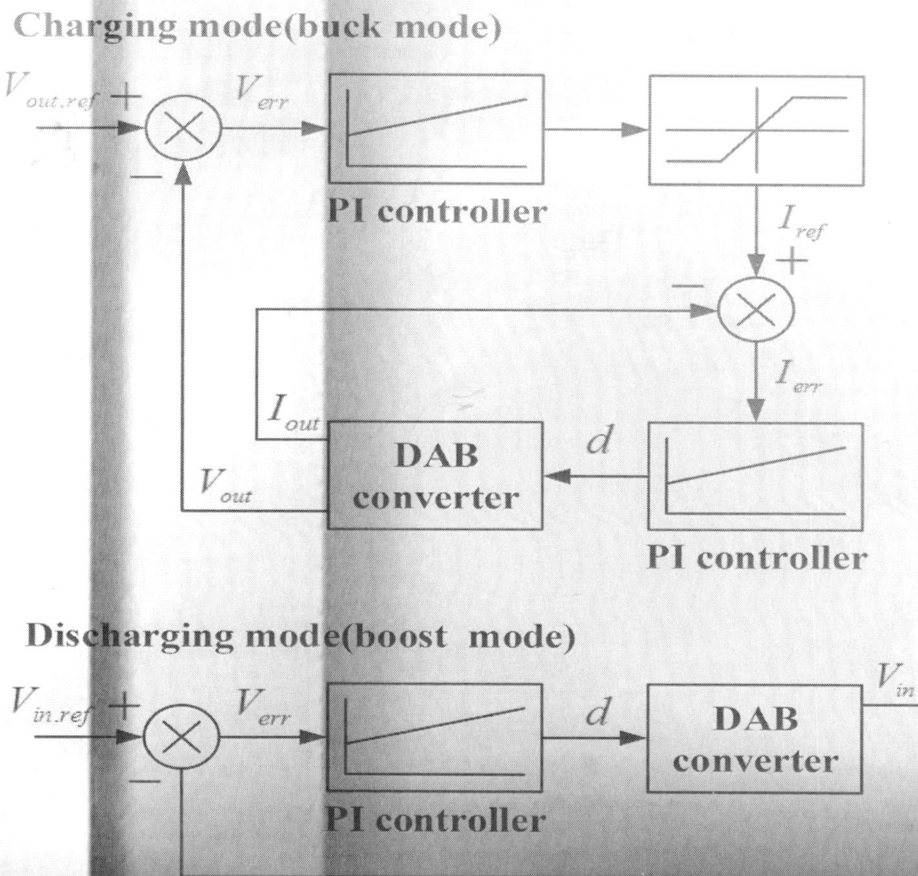

Solar Power and Energy Storage Systems

Solar Power and Energy Storage Systems

Hee-Je Kim

PAN STANFORD PUBLISHING

Published by

Pan Stanford Publishing Pte. Ltd.
Penthouse Level, Suntec Tower 3
8 Temasek Boulevard
Singapore 038988

Email: editorial@panstanford.com
Web: www.panstanford.com

British Library Cataloguing-in-Publication Data
A catalogue record for this book is available from the British Library.

Solar Power and Energy Storage Systems

ISBN 978-981-4800-13-6 (Hardcover)
ISBN 978-0-429-45877-4 (eBook)

Contents

Preface

Today's energy path depends largely on fossil fuels, and we are already facing alarming consequences in terms of climate change and energy security. Solar cells have been recognized as an important alternative power source since the 1970s. Solar cells are also promising as a carbon-free energy source that can suppress global warming. The power conversion efficiency of a solar cell is well defined as the ratio between the electric power produced by the solar cell and the incident sunlight energy per unit of time. At present the highest reported cell efficiencies in laboratories are around 40%, while the power conversion efficiencies for thermal power generation can exceed 50%. This, however, by no means indicates any advantage related to thermal generation, since its resources, such as fossil fuels, are limited while solar energy is fundamentally unlimited. However, during nighttimes, we cannot get electricity from the sun, so we need some storage systems for nighttimes.

Design and fabrication of electrochemical energy storage systems with both high energy and power densities as well as long cycling life is of great importance. As one of these systems, a battery-supercapacitor hybrid device (BSH) is typically constructed with a high-capacity battery-type electrode and a high-rate capacitive electrode, which has attracted enormous attention due to its potential applications in, for example, future electric vehicles, smart electric grids, and even miniaturized electronic/optoelectronic devices. Before we start research on BSH, we must study supercapacitors. So I have introduced the results of several supercapacitor studies in Chapters 5 and 6.

In this book I have introduced several examples of solar energy applications, such as solar ships, DAB converters, and transformerless inverters. In addition, I have described how to increase the efficiency of conversion of sunlight to electricity for the future. In the last chapter, Chapter 7, remote control by using a personal smartphone is described.

Chapter 1

Hybrid Photovoltaic/Diesel Green Ship Operating in Stand-Alone and Grid-Connected Modes: Experimental Investigation

In this chapter, we describe the experimental results from the operation of a prototype green ship in Geoje island, South Korea. As Korea is a peninsula situated between the continent and the sea, its location is ideal for utilizing a photovoltaic (PV)/diesel green ship. After ground testing with a stand-alone PV generation system, this PV system was added to a conventional diesel ship. The prototype green ship consisted of a PV generation system, a diesel engine, battery energy storage, a hybrid control system, and a stand-alone and grid-connected inverter. The aim of the green ship was not only to minimize fuel consumption but also to support the power grid as distributed generation connected to the smart grid (on land) and microgrid (on island) in the near future. At the end of this paper, environmental, economic, and sensitivity analysis mainly based on experimental results will be discussed.

Solar Power and Energy Storage Systems
Hee-Je Kim

ISBN 978-981-4800-13-6 (Hardcover), 978-0-429-45877-4 (eBook)
www.panstanford.com

1.1 Overall Survey

Around the world, governments have strived to increase the share of renewable green energy in power production. Study of solar energy is increasing because of the serious global warming. Conversion of sunlight into electricity is very important as an alternative source. Solar energy is considered as the best source of renewable energy, and it is unlimited clean energy. Solar radiation can be harnessed and converted into different forms of energy, and it will not pollute the environment. To transform solar radiation we need collectors of sunlight, like solar cells. The main interests have been energy security, the increasing prices of carbon-based energy sources, and minimization of global warming. Global shipping is a major contributor to global greenhouse gas (GHG) emissions, being responsible for approximately 3% of global CO_2 emissions.[1,2] The International Maritime Organization is now working to start GHG regulations for global shipping and is under pressure, for example, from the EU and the United Nations Framework Convention on Climate Change, to apply regulations that will have a substantial impact on emissions.[3] In addition, because global shipping is normally powered by stand-alone diesel generators for electricity supply, the shipping industry is affected considerably by the increase in global fuel prices. Therefore, the use of a renewable energy resource in the shipping industry would be a great advantage, particularly in reducing CO_2 emissions and the dependence on the highly unpredictable diesel price.[4] The mandatory renewable energy targets in Korea are expected to be 11% in 2020 compared to 2.4% in 2008, as shown in Fig. 1.1, which shows increasing trends in the world energy policy.[23] Therefore, every possible method for increasing the proportion of renewable energy is required to meet the goals and expectations. Many studies have examined hybrid energy systems.[5–15] Among the renewable energy sources, the abundant solar energy is most applicable to conventional ships at the sea.[16,17] Abdullaha et al.[24] proved that hybrid power schemes are more sustainable in terms of supplying electricity in rural areas compared to a stand-alone photovoltaic (PV) system due to prolonged cloudy and dense haze periods. The investigators

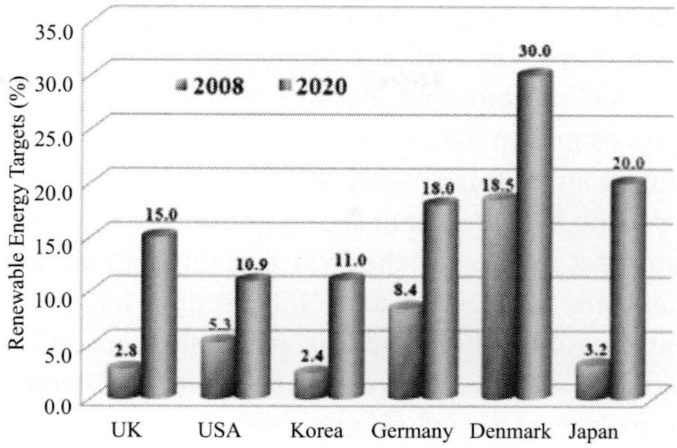

Figure 1.1 Renewable energy targets (%).

presented reliable and cost-effective hybrid energy systems.[25] A hybrid green ship (hybrid PV/diesel) might be an effective solution.

This chapter proposes a hybrid PV/diesel green ship operating in a stand-alone and grid-connected modes. PV is the primary power source of the system for the electrical loads in the ship, and the diesel engines are used to compensate for the fluctuating electric power from the PV system. A battery bank is also installed in the hybrid system to supply stable power.[18] Unlike a conventional PV hybrid system on land, a hybrid green ship utilizes the free space on the ship. In addition, with grid-connected inverters, the overall power flow is broadened from the smart grid on land to the off-grid on the island, which is more economical than a hybrid PV (stand-alone only)/diesel system. Also, Korea is a peninsula that stretches from the north to the south, located at latitude $33^\circ \approx 43^\circ$N, longitude $124^\circ \approx 132^\circ$E. As Korea is situated between the continent and the sea, its location is ideal for utilizing a PV/diesel green ship between the continent and the sea. Therefore, the proposed hybrid PV/diesel green ship is effective in countries with these territory characteristics.

This chapter reports the experimental results from the operation of a prototype green ship in Geoje island, South Korea. After ground testing with a stand-alone PV generation system, the PV system was adopted for a conventional diesel ship. The prototype green ship consisted of a PV generation system, a diesel engine, battery energy storage, a hybrid control system, and a stand-alone and grid-connected inverter. The aim of the green ship was not only to minimize fuel consumption but also to support the power grid through a smart grid in the near future.

The remainder of this chapter is organized as follows: Section 1.2 discusses the hybrid system configuration and operation mode. The stability test results for the hybrid green ship equipped with a PV system are given in Section 1.3. Section 1.4 discusses the experimental results of the hybrid green ship according to the operation mode. Section 1.5 gives an environmental, economic, and sensitivity analysis of the hybrid green ship.

1.2 Proposed Hybrid PV/Diesel Green Ship Experimental System

The operation of the hybrid PV/diesel green ship is divided mainly into modes A (at sea) and B (in port). The power ratings of the hybrid PV/diesel system were approximately 3.2 kW for the electrical loads, such as lighting, global positioning system (GPS), and communication system and video system, in the cruise ship (85 ton). Figure 1.2 shows a block diagram of the proposed hybrid system. Table 1.1 provides details of the hybrid system components. In the conventional ship, two diesel-powered generator (20 kW) are installed to meet the occasional peak load of the winch motor to pull the ship anchor. The power rating of base electrical loads of the ship is under 3 kW, which can be thoroughly covered by the PV system.

1.2.1 Power Conversion System in a Hybrid Green Ship

As mentioned above, there are mainly two operation modes. Mode A (at sea) is like the stand-alone mode of PV systems on land, as shown in Fig. 1.2. According to the status of PV power, the electrical loads in ships can be powered by either a battery bank or a diesel generator. The PV arrays on the left and right sides of the roof are connected separately to a solar controller (DC–DC converter) due to the different maximum power point (MPP). Once the cruise ship is in harbor and the battery is fully charged, the extra PV power

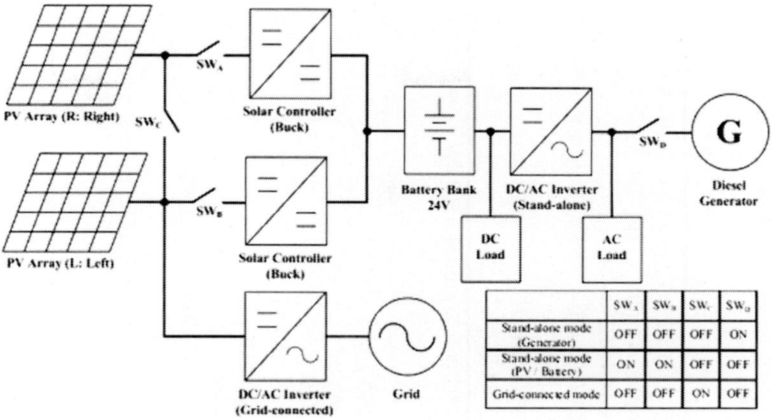

Figure 1.2 Block diagram of a hybrid PV/diesel green ship.

can be utilized through a grid-connected inverter. To meet the input voltage range of the grid-connected inverter, the PV arrays on the left and right are connected in series, not requiring an additional power conversion unit, resulting in a doubling of the V_{OC} (open circuit voltage). The control of the hybrid systems is summarized in Fig. 1.2, where SW_A, SW_B, and SW_C are the magnet contact switches controlled by a laboratory-made hybrid control panel. The auto transfer switch (ATS) is installed between the output of the DC–AC inverter (stand-alone) and the output of the diesel generator to simply change the operation mode.

The monitoring system for the hybrid green ship was established using the LabVIEW™ program, as shown in Fig. 1.3. The monitoring system was connected to the PC through an RS232 communication, storing the acquired data in the Microsoft Excel file. The controller board includes an Atmel AVR ATmega 128. The prototype hybrid green ship plied between Geojedo (island) and Somaemuldo (island). Through the experiment, the validity of the hybrid green ship was confirmed.

1.2.2 Implementation of the Hybrid Green Ship

After the verification experiment of the hybrid PV/diesel system on land in 2010, the proposed system was installed as an 85-ton level of

Table 1.1 System component parameters

PV array	Module unit	200 W under 1 sun (100 mW/ cm^2), 25°C V_{oc} = 33.2 V, I_{sc} = 8.09 A	
	Module number	4(Ns) × 2(Np) × 2(group) = 16	
	Power rating	200 × 16 = 3.2 kW	
Battery	Cell unit	12 V/100 AH	
	Cell number	2(Ns) × 8(Np) = 16	
	Capacity	19.2 kWh	
	Rated voltage	24 V	
Power conversion system (MSB, main switch board)	DC–DC converter (solar controller)	Power rating	1.6 kW × 2(group) = 3.2 kW
		Maximum solar input	150 VDC
		Output	24 VDC
	DC–AC inverter (stand-alone)	Power rating	4 kW
		Input	22.5–30.8 VDC
		Output	220 VAC
	DC–AC inverter (grid connected)	Power rating	4 kW
		Input	100–370 VDC
		Output	220 VAC
	NFB	2P-20A, 2P-30A, 2P-175A	
	MC (magnet contact)	MC-32	
	ATS (auto transfer switch)	DMH20	

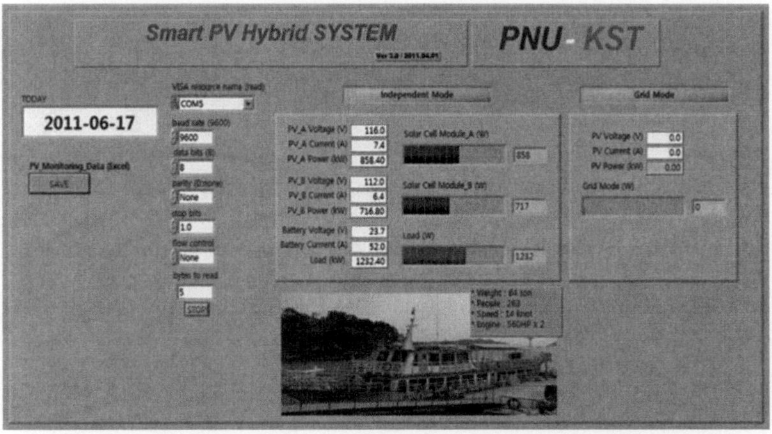

Figure 1.3 Monitoring the system for a hybrid green ship (LabVIEW™ program).

the cruise ship in 2011, in Geoje island. After searching for a suitable ship, a conventional ship was renovated according to the proposed design mentioned above.

Unlike the PV plants on land, the total weight of the PV generation system and the wind pressure on the PV arrays are the main concerns regarding the safety of a ship at sea. Approved by the Korea Ship Safety Technology Authority (KST), PV systems were installed on a conventional cruise ship and controlled by a hybrid control panel. 3.2 kW (200 W/16 EA) of PV arrays were installed at the back of the ship, as shown in Fig. 1.4. A battery bank was installed in the basement of the ship, taking into consideration the stability of the ship. Section 1.3 provides details of the safety problems.

1.3 Stability of a Hybrid Green Ship

The selected ship for the experiment was a normal passenger ship (85 ton, 263 passengers). Stability is one of the greatest concerns for a passenger ship. The back of the dock was renovated according to the design, as shown in Fig. 1.5, considering the mean height of the

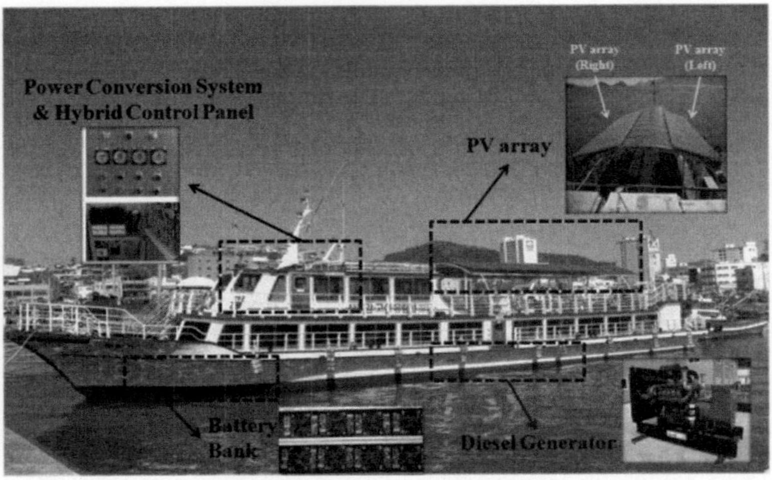

Figure 1.4 Prototype of a hybrid PV/diesel green ship (Geoje Island, South Korea/34.7269°N, 128.6042°E).

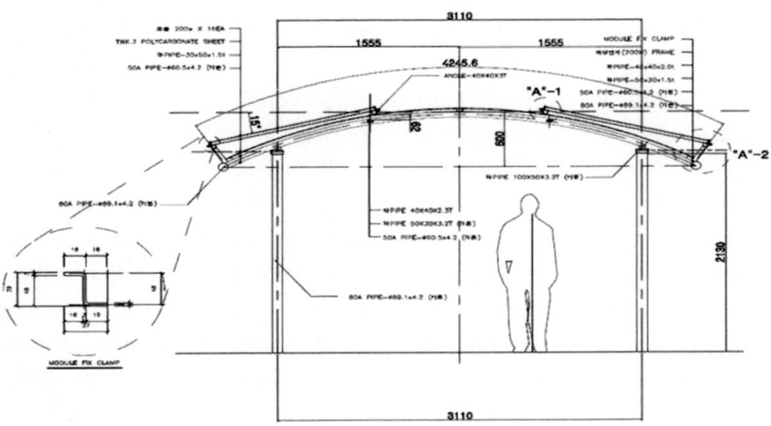

Figure 1.5 Installation drawing of PV arrays.

passengers, wind pressure, and maintenance. During the verification experiment of the hybrid PV/diesel system on land in 2010, the power production of the PV system was measured by changing the tilt angle of the solar panel. The results showed that the group with a 0° tilt showed poor power production due to dirt on the solar panels. In this study, a 15°-tilted solar panel was installed instead of the awning of a conventional ship. A polycarbonate sheet, which is a transparent plastic material with outstanding impact resistance, was also installed under the PV arrays for passenger safety. To evaluate the stability of the hybrid green ship, the weight of each component was checked and assessed by the KST. The main objective of the KST is to inspect vessels in accordance with the Ship Safety Act to prevent exposing the vessel and crew to any unpredictable dangers that may occur at sea.

According to Article 28 (Keeping Stability) of the Ship Safety Act in Korea, the owner of a ship that falls under any of the following subsections shall maintain stability according to the criteria determined/publicly announced by the minister in the event a ship, such as a passenger ship, is determined by the Ordinance of the Ministry of Land, Transport and Maritime Affairs to be 12 m or longer. In connection with stability under Article 28, an owner of a ship shall submit the materials for stability to the minister for approval as to whether or not the stability is appropriate and shall provide all the necessary materials approved for stability to the captain of the ship.[19]

In this study, a conventional ship was modified to a hybrid green ship without changing the main frame size of the ship. The KG (vertical center of gravity) and the LCG (longitudinal center of gravity) were affected slightly by the removal and installation of components, such as PV arrays, power conversion units, and polycarbonate sheets. An authorized inspector checked the renovated hybrid green ship and identified an increase in the LIGHT WEIGHT (approximately 1.154 tons, a 1.38% increase) and displacement of the KG (a 0.033 m increase) and LCG (0.029 m to the bow of the ship). The renovated hybrid green ship passed the stability inspection by the KST, as shown in Table 1.2, showing a displacement of LIGHT WEIGHT between 0.5% and 2%, with an increase in KG. In the proposed green ship, the total increase in weight is 1.154 ton, which is 1.3% of the total weight of the ship and 7% of the total weight of the maximum number of passengers with the average weight of an adult human (62 kg), which did not impact energy saving potential.

Table 1.2 Stability assessment of the hybrid green ship (LCG and KG)

Inspection sheet					
1. New light calculation					
Item	Weight [t]	LCG [m]	Moment [T m]	KG [m]	Moment [T m]
Old light	83.333	−1.224	−101.936	1.513	126.091
Deduction	0.032	−6.700	−0.214	6.150	0.197
Addition	1.186	0.672	0.797	3.996	4.739
New light	84.487	−1.195	−100.925	1.546	130.633
2. New LIGHT WEIGHT and KG displacement					
(1) LIGHT WEIGHT displacement:		1.154	[t] Increase (approximately 1.38%)		
(2) KG displacement:		0.033	[m] Increase		
(3) LCG displacement:		0.029	[m] Increase to the bow of the ship		
3. Result (under the Ship Safety Act)					
KG (increase)	The displacement of LIGHT WEIGHT (W, %)	$W < 0.5$	Exemption of stability test		
		$0.5 \leq W < 2$	Ship stability calculation with (displaced LCG, original KG value)		
		$W \geq 2$	Retest		

KG (decrease)	The displacement of LIGHT WEIGHT (W, %)	$W < 1.5$	Exemption of stability test
		$1.5 \leq W < 2$	Ship stability calculation with (displaced LCG, original KG value)
		$W \geq 2$	Retest

*Decision: evaluated LIGHT WEIGHT
(KG increased, the displacement of LIGHT WEIGHT (W, %) is between 0.5% and 2%.

Deduction item					
Item	Weight [t]	LCG [m]	Moment [T m]	KG [m]	Moment [T m]
Tent	0.032	−6.700	−0.214	6.150	0.197
Total	0.032	−6.700	−0.214	6.150	0.197
Addition item					
Item	Weight [t]	LCG [m]	Moment [T m]	KG [m]	Moment [T m]
Steel structure	0.187	−6.700	−1.253	6.250	1.169
PV arrays	0.306	−6.700	−2.050	6.250	1.913
Polycarbonate sheets	0.143	−6.700	−0.958	6.250	0.894
Battery bank	0.400	9.750	3.900	0.500	0.200
Inverter, converter	0.028	9.750	0.273	0.500	0.014
Control panel	0.122	7.250	0.885	4.500	0.549
Total	1.186	0.672	0.797	3.996	4.739

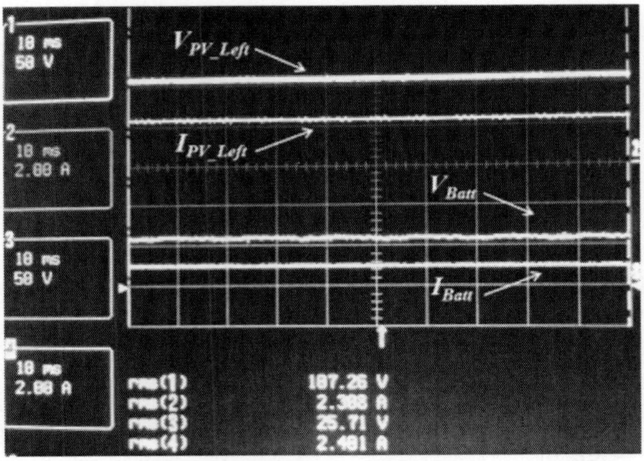

Figure 1.6 Experimental waveforms at PV arrays.

1.4 Experimental Results of a Hybrid PV/Diesel Green Ship

The experimental results presented in this section were recorded after an inspection by the KST and obtaining permission for ship operation by the Korean Maritime Police. The experiment was conducted according to the operation mode, as mentioned in Section 1.2.

First, in the stand-alone mode, the outputs of the PV arrays and the battery bank were checked with only the DC loads turned on, such as lighting, communication, and GPS equipment. Figure 1.6 presents the PV array voltage of the left side ($V_{\text{PV\_Left}}$), the PV array current of the left side ($I_{\text{PV\_Left}}$), the battery voltage (V_{Batt}), and the battery current (I_{Batt}) to the DC loads. In Fig. 1.5, $V_{\text{PV\_Left}} = 107.26$ V and $I_{\text{PV\_Left}} = 2.38$ A, resulting in a power of 255 W to the solar controller. In addition, the battery voltage was 25.7 V, which is equal to the output of the solar controller, and the battery current was 2.401 A during charging. In this case, the power to the loads was less than the power from the PV arrays.

Figure 1.7 shows the battery voltage (V_{Batt}), battery current (I_{Batt}), voltage of the stand-alone inverter (v_{INV}), and current of the stand-alone inverter (i_{INV}) to electrical loads in the operation of the

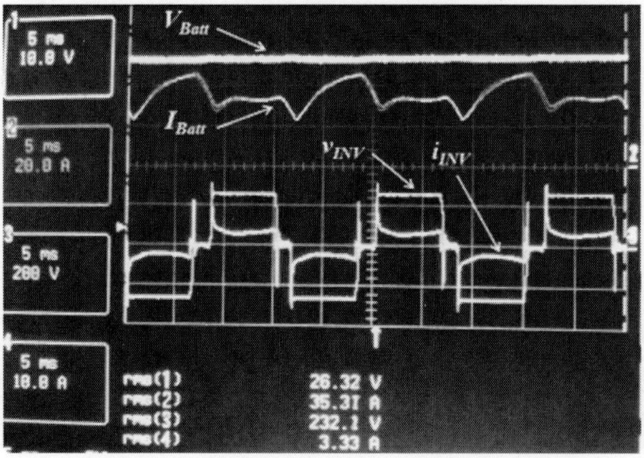

Figure 1.7 Experimental waveforms in the stand-alone mode (before the cruise).

passenger ship between Geojedo (island) and Somaemuldo (island). Unlike Fig. 1.6, during the course, the electrical loads included the DC and AC loads. In the experiment, a pulse width modulation–modified sine wave type was used for the stand-alone mode inverter, as shown in Fig. 1.7, resulting in a square waveform voltage and current with 60 Hz. During the measurement, the power supplied to the loads was approximately 773 W.

Figure 1.8 shows the logged data of PV production and power to the loads during the cruise. The passenger ship began to cruise at 10:24 a.m. (July 21, 2011) from Geojedo (island) to Somaemuldo (island). There was a difference in power output between the PV array of the left side and the PV array of the right side. This is because the PV arrays were tilted 15° (see Fig. 1.4) and the passenger ship was affected by the rolling, pitching, and yawing movements.

On the other hand, to solve the problem of the different MPP, the solar controllers were connected separately to the PV arrays on each side, tracking the MPP per arrays. The electrical loads were varied arbitrarily by the captain of the ship. When the output power of the PV arrays was higher than the electrical loads, the battery bank was charged by the excess amount for later use. On the contrary, the battery bank supported the shortage when the output power from the PV arrays was less than the connected electrical loads.

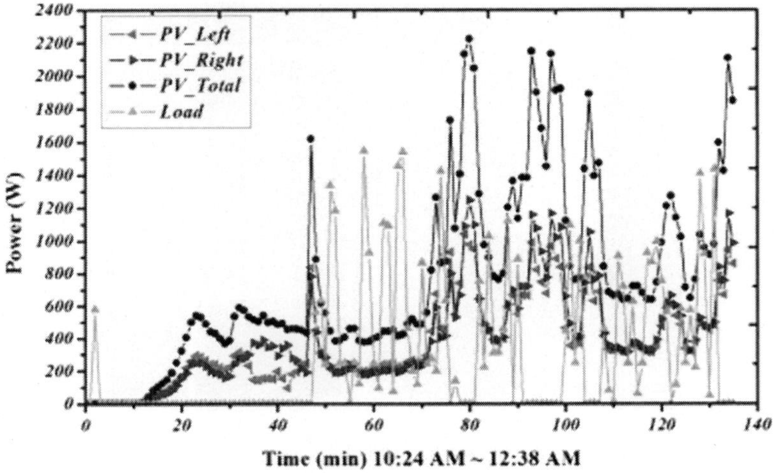

Figure 1.8 Experimental waveforms in the stand-alone mode (during the cruise).

In this study, a grid-connected mode was implemented when the ship was at anchor in the harbor. The PV arrays were connected in series using a simple change in magnet switches, resulting in N_s (number of series connections) = 8 and N_p (number of parallel connections) = 2. The open-circuit voltage of the PV arrays in the grid-connected mode was 265.6 V, which satisfies the input voltage range of a DC–AC inverter (grid connected). In the case of grid faults, such as grid over/undervoltage, grid over/underfrequency, overcurrent, and ground fault, the grid-connected inverter will wait 5 min. prior to restarting. Figure 1.9 shows the voltage of the PV arrays (V_{PV_Total}), current of the PV arrays (I_{PV_Total}), AC grid voltage (v_{Grid_INV}), and AC grid current (i_{Grid_INV}) at Geojedo (island). The output voltage of the PV arrays in the grid-connected mode was increased to 224 V, which is the voltage at MPP (V_{mpp}). The injected active power to the grid was approximately 600 W at 3:20 p.m. (Sep. 27, 2011), as shown in Fig. 1.9. The injected grid current includes low-order harmonic components due to the effect of the weak and distorted grid at Geojedo (island), requiring an improvement in power quality using a shunt active power filter in the near future.

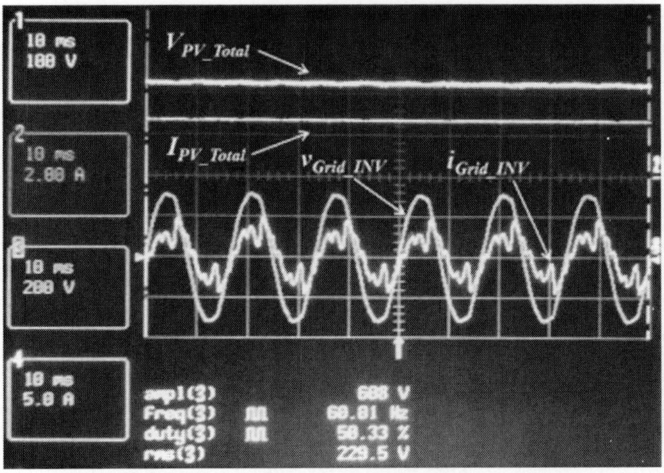

Figure 1.9 Experimental waveforms in the grid-connected mode.

The performance of the hybrid PV/diesel operation was evaluated by changing the electrical loads arbitrarily when the ship was at anchor in the harbor (Oct. 31, 2011), as shown in Fig. 1.10. During the performance test, a change in operation mode occurred three times from PV mode to diesel mode and vice versa. It takes 8–10 s for the diesel generator to reach normal operation. However, the power from the battery bank supported the transient stability, including critical loads, such as a GPS navigator, light detection and ranging, and lighting.

1.5 Environmental and Economic Analysis of the Proposed Hybrid PV/Diesel Green Ship

The Korean government has introduced a number of new incentives and support programs, such as the "1 million green homes" program. Korea's electricity prices are cheaper than in Europe and Japan, but they are likely to rise in the near future. This will trigger the installation of PV systems on rooftops. On the other hand, these support programs for PV systems are limited to the ground-mount type or rooftop type and a lack of space will be a severe limitation for

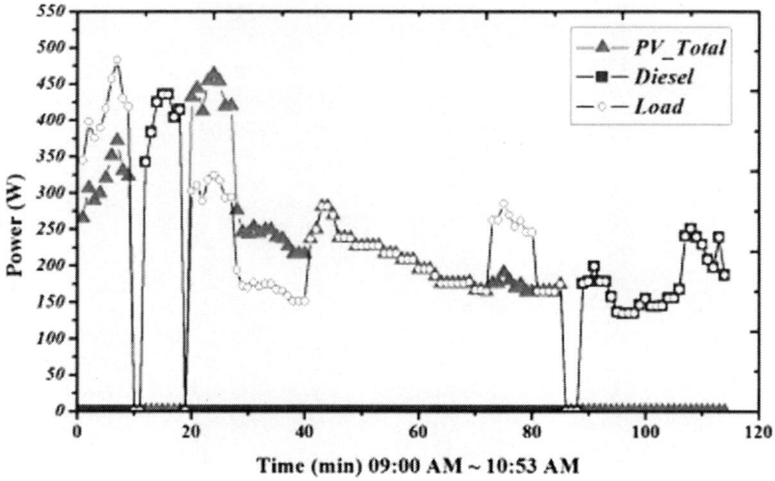

Figure 1.10 Power output of the hybrid PV/diesel systems.

PV installation. Therefore, the hybrid PV/diesel green ship can be an alternative solution, particularly in countries with small territorial dimensions. In addition, remote areas, such as islands, are greatly affected by the increase in global fuel prices.[4] Consequently, a hybrid PV/diesel green ship would be of great benefit in these locations.

1.5.1 Environmental Analysis

The proposed hybrid PV/diesel green ship could reduce 3.5 tons of GHG equivalent of CO_2 annually. In addition, a total of 70 tons of GHG could be avoided during the lifetime of 20 years. The reduction of GHG on 70% PV penetration was compared with a diesel-only powered ship, given in Table 1.3. The reduction in CO_2 with increasing PV penetration of 30%, 50%, and 70% is depicted in Fig. 1.11, which shows a positive impact on CO_2 and fuel savings potential of the green ship.

1.5.2 Economic and Sensitivity Analysis

The total cost of each component of the hybrid PV/diesel green ship, including mainly the PV module, the solar battery bank, and the power conversion unit, is shown in Table 1.4.

Table 1.3 Annual CO_2 emissions and fuel savings for a hybrid PV/diesel green ship

Diesel only	70% PV penetration	
Operating days (annually)	150	150
Operating hours of the diesel engine (daily)	6	1.8
Diesel consumption rate (L/h)	3	
Diesel (DM grade): density (kg/m^3)	830	
Diesel (DM grade): % of carbon	85	
Emission of carbon (ton)	1.905	0.571
Carbon to carbon dioxide (44/12)	3.667	
Emission of carbon dioxide (ton)	6.984	2.095
Avoided emission of carbon dioxide (ton/year)	–	4.889
Fuel saving (kL)	–	1.6

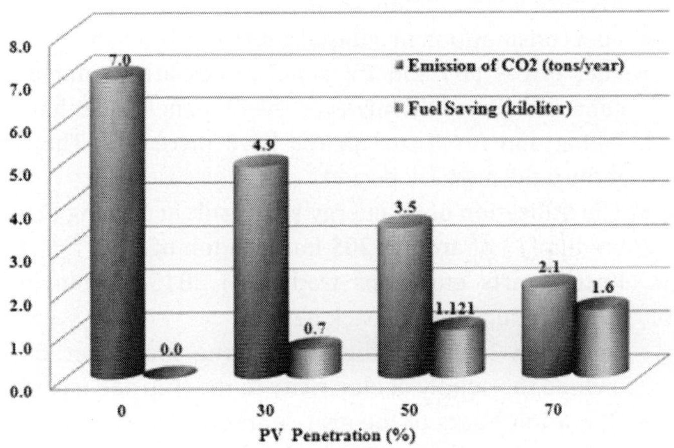

Figure 1.11 Emission of CO_2 and fuel saving potentials according to the PV penetration (%).

An economic analysis has been performed in this section to encourage the installation of the hybrid PV/diesel system on a ship. To simplify the cost analysis, the following principles and equations were used to calculate the payback period:

- The initial investment cost (A) includes the components of the PV systems and the renovation construction cost in a conventional ship.

Table 1.4 Cost analysis of the hybrid PV/diesel green ship

Component	Unit price ($)	EA	Price ($)
V module (200 W)	660	16	10,560
Solar controller	800	2	1600
Solar battery	100	16	1600
DC–AC inverter (stand-alone)	1000	1	1000
DC–AC inverter (grid connected)	1600	1	1600
Hybrid control panel	3000	1	3000
Construction cost (frame)	8000	1	8000
Construction cost (electrical works)	4000	1	4000
Initial investment cost (A)			31,360

- The maintenance cost is not involved in the calculation because it is negligibly small in PV systems.
- Fuel consumption in a diesel generator is 3 L/h.
- Fuel prices (B) and PV panel prices are assumed to be constant. (Recently, however, the PV panel prices have been falling and fossil fuel prices have increased. This would reduce the payback period.)
- The utilization of PV energy will result in earning of carbon credits (J) of around 20$ for each ton of GHG.[20, 21] (South Korea starts emissions trading in 2015. Carbon pricing would reduce the payback period.)
- Operation hours of the stand-alone mode per day (E) include the supply of electricity to the ship and island.
- Operation hours of the grid-connected mode per year (F) include the supply of electricity to the power grid. The electricity bill (G) is assumed to be 0.1$/kWh in Korea.
- The maximum number of operation days of the hybrid PV/diesel system is set to 150 days/year. The maximum number of operation hours of the hybrid PV/diesel system is set to 6 h/day ($= E + F$).
- There will be no financial support from the government, such as tax incentives. (This factor would reduce the payback period.)
- The loan rate[22] offered from the lending institution for the initial investment cost is not considered.

Table 1.5 The cost analysis of the hybrid PV/diesel green ship

Percentage of PV penetration	70%
Initial investment cost (A)	31,360$
Fuel price (B)	1.7$/L
Diesel consumption rate (C)	3 L/h
Operation of hybrid PV/diesel green ship (D)	150 days
Operation hours of the stand-alone mode (E)	3 h/day
Operation hours of the grid-connected mode (F)	1.2 h/day
Electricity bill (G)	0.1$/kWh
Power rating of the PV system (H)	3.2 kW
Total efficiency of the PV system (I)	0.94
Carbon credit (J)	20$/ton
Avoided emission of carbon dioxide (K)	4.889 ton/year
Payback period (X)	11.87 years

According to these principles, the payback period was calculated using the following equation:

$$X = AD \times (B \times C \times E + F \times G \times H \times I) + J \times K \qquad (1.1)$$

The payback period of the proposed hybrid PV/diesel green ship is 11.87 years according to the assumptions in Table 1.5.

The payback period decreases with an increase in the number of operating hours of the stand-alone mode and the possible operation days of the PV/diesel green ship. Also, the stand-alone and grid-connected modes (including the operation on land and island) within 6 h of the maximum operation hours per day show potential supply of electricity to the power grid, as shown in Fig. 1.12. When the electricity bill is increased, the payback period of the hybrid PV/diesel green ship is reduced, which is the denominator part of the equation above. South Korean electricity prices remain much lower than those in other developed countries, as shown in Fig. 1.13. South Korea raised electricity prices twice in 2012 to help the country's monopoly power distributor to cover the higher fuel costs.[26]

The payback period is expected to be reduced more due to the decreasing PV module costs, increasing diesel fuel costs, and CO_2 emission regulations.[20–22]

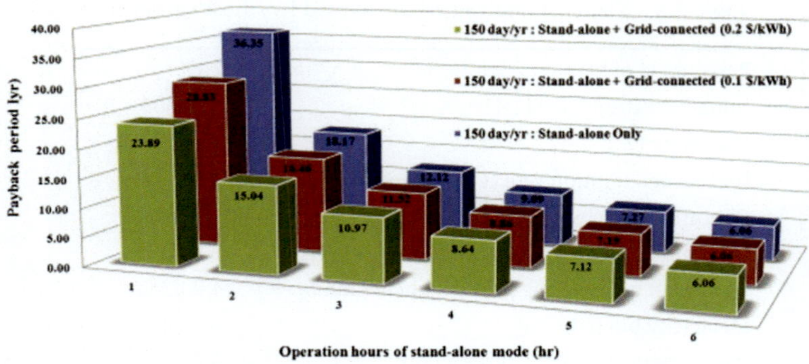

Figure 1.12 Payback period according to the operation hours of the hybrid PV/diesel green ship and the electricity price.

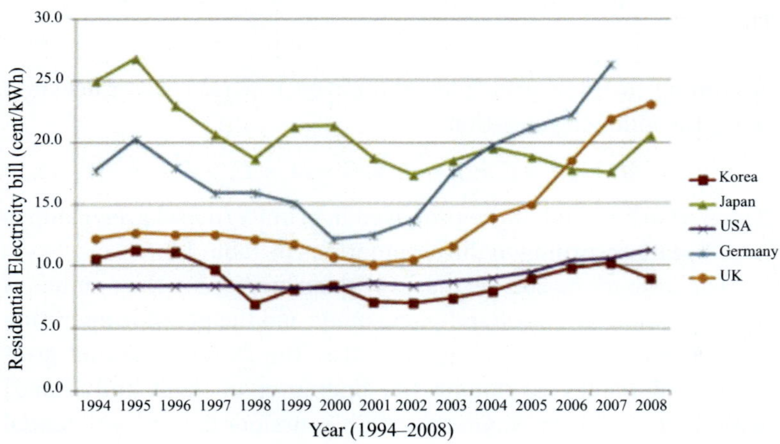

Figure 1.13 Residential electricity bill of Korea (1994–2008).

The choice of the hybrid PV/diesel green ship is mainly dependent on the diesel price, being feasible if the price is more than 1.05$/L.[4] As the diesel price on the market in Korea is already above 1.05$/L, as shown in Fig. 1.14, the hybrid PV/diesel green ship is not expected to be sensitive to variation in the diesel price.[27] Although the hybrid PV/diesel green ship will ultimately be cost effective, a government subsidy is still essential for its widespread use.

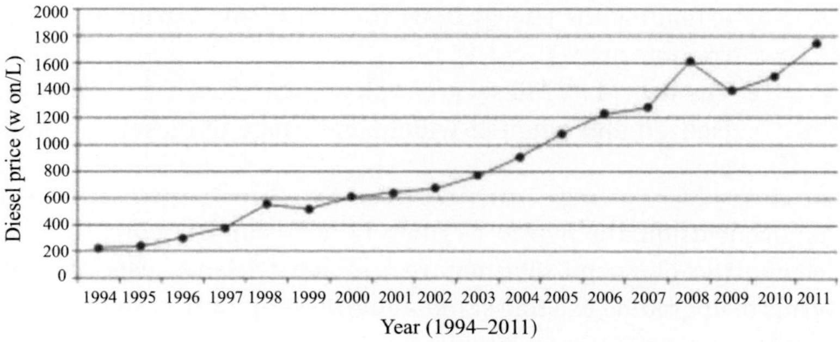

Figure 1.14 Diesel price in Korea (1994–2011).

1.6 Conclusions

This chapter proposed that the hybrid PV/diesel green ship operates not only in a stand-alone mode but also when connected to a smart grid. To commercialize the proposed green ship in the near future, a conventional passenger ship was fitted with a 3.2 kW PV system and operated during the project. The operating strategy of the hybrid PV/diesel system, stability assessment, and economic analysis were discussed and the following conclusions were made:

- 3.2 kW PV arrays were fixed on the top of the green ship with a tilt of 15°, taking into consideration the cost of installation, maintenance cost, and stability against wind.
- The stand-alone mode and the grid-connected mode were controlled automatically or manually by the captain of the ship.
- The hybrid PV/diesel operation was stable with the battery bank during mode changes.
- The hybrid PV/diesel green ship passed the stability assessment test by the authorized department (KST in this project).
- The payback period of the initial investment can be reduced due to the decreasing PV module costs, increasing diesel fuel costs, and CO_2 emission trading system.
- The hybrid PV/diesel green ship will decrease the dependence on fossil fuels and might be one solution for

extending the energy flow from land (smart grid) to island (remote areas).

- The hybrid PV/diesel green ship is expected to be commercialized and adopted widely in a range of diesel-powered ships.

As a conclusion, the hybrid PV/diesel green ship has potential use in countries of a peninsular nature or of many islands, especially in terms of upgrading existing stand-alone diesel-powered ships under 100 tons due to GHG regulations.

References

1. Ø. Buhaug, J. J. Corbett, Ø. Endresen, V. Eyring, J. Faber, S. Hanayama, et al. (2009). Second IMO GHG study 2009, International Maritime Organization (IMO), London, UK.
2. Ø. Endresen, M. Eide, S. Dalsøren, I. S. Isaksen, E. Sørgard (2008). The environmental impacts of increased international maritime shipping, past trends and future perspectives, OECD/ITF Global Forum on Transport and Environment in a Globalising World, Guadalajara, Mexico.
3. K. Van Dender, P. Crist (2008). Policy instruments to limit negative environmental impacts from increased international transport: an economic perspective, Global Forum on Transport and Environment in a Globalising World, Guadalajara, Mexico.
4. K. Y. Lau, M. F. M. Yousof, S. N. M. Arshad, M. Anwari, A. H. M. Yatim (2010). Performance analysis of hybrid photovoltaic/diesel energy system under Malaysian conditions, *Energy*, **35**, pp. 3245–3255.
5. Ø. Ulleberg, S. O. Mørner (1997). TRNSYS simulation models for solar hydrogen systems, *Sol. Energy*, **59**(4–6), pp. 271–279.
6. R. Chedid, H. Akiki, S. Rahman (1998). A decision support technique for the design of hybrid solar-wind power systems, *IEEE Trans. Energy Convers.*, **13**(1), pp. 76–83.
7. W. D. Kellogg, M. H. Nehrir, G. Venkataramanan, V. Gerez (1998). Generation unit sizing and cost analysis for stand-alone wind, photovoltaic, and hybrid wind/PV systems, *IEEE Trans. Energy Convers.*, **13**(1), pp. 70–75.
8. E. S. Abdin, A. M. Osheiba, M. M. Khater (1999). Modeling and optimal controllers design for a stand-alone photovoltaic-diesel generating unit, *IEEE Trans. Energy Convers.*, **14**(3), pp. 560–565.

9. F. Bonanno, A. Consoli, A. Raciti, B. Morgana, U. Nocera (1999). Transient analysis of integrated diesel-wind-photovoltaic generation systems, *IEEE Trans. Energy Convers.*, **14**(2), pp. 232–238.

10. H. Sharma, S. Islam, T. Pryor (2000). Dynamic modeling and simulation of a hybrid wind diesel remote area power system, *Int. J. Renewable Energy Eng.*, **2**(1), pp. 19–25.

11. F. Giraud, Z. M. Salameh (2001). Steady-state performance of a grid-connected rooftop hybrid wind-photovoltaic power system with battery storage, *IEEE Trans. Energy Convers.*, **16**(1), pp. 1–7.

12. Th. F. El-Shatter, M. N. Eskandar, M. T. El-Hagry (2002). Hybrid PV/fuel cell system design and simulation, *Renewable Energy*, **27**(3), pp. 479–485.

13. M. T. Iqbal (2003). Modeling and control of a wind fuel cell hybrid energy system, *Renewable Energy*, **28**(2), pp. 223–237.

14. K. Agbossou, M. Kolhe, J. Hamelin, T. K. Bose (2004). Performance of a stand-alone renewable energy system based on energy storage as hydrogen, *IEEE Trans. Energy Convers.*, **19**(3), pp. 633–640.

15. D. B. Nelson, M. H. Nehrir, C. Wang (2006). Unit sizing and cost analysis of stand-alone hybrid wind/PV/fuel cell systems, *Renewable Energy*, **31**(10), pp. 1641–1656.

16. J. H. R. Enslin, D. B. Snyman (1991). Combined low-cost, high-efficient inverter, peak power tracker and regulator for PV applications, *IEEE Trans. Power Electron.*, **6**(1), pp. 73–82.

17. S. Jain, V. Agarwal (2004). A new algorithm for rapid tracking of approximate maximum power point in photovoltaic systems, *IEEE Power Electron. Lett.*, **2**(1), pp. 16–19.

18. C. Wang, M. H. Nehrir (2008). Power management of a stand-alone wind photovoltaic fuel-cell energy system, *IEEE Trans. Energy Convers.*, **23**(3), pp. 957–967.

19. Ship Safety Act (2009). Enforcement date 06 Feb., 2009, Act no. 9446, Partial Amendment, http://law.go.kr

20. S. Rehman, L. Al-Hadhrami (2010). Study of a solar PV-diesel-battery hybrid power system for a remotely located population near Rafha, Saudi Arabia, *Energy*, **35**, pp. 4986–4995.

21. S. K. Nandi, H. R. Ghosh (2010). Prospect of wind-PV-battery hybrid power system as an alternative to grid extension in Bangladesh, *Energy*, **35**, pp. 3040–3047.

22. Z. Li, F. Boyle, A. Reynolds (2011). Domestic application of solar PV systems in Ireland: the reality of their economic viability, *Energy*, **36**, pp. 5865–5876.

23. Korea Energy Economics Institute (KEEI) [Internet]. Available from: <http://www.keei.re.kr/web_keei/d_results.nsf/0/A33A41DF8C0CF20B4925777D00192DA7/$file/EC%97%B0%EC%9E%9010-01.pdf>.
24. M. O. Abdullaha, V. C. Yunga, M. Anyia, A. K. Othmana, K. B. A. Hamida, J. Taraweb (2010). Review and comparison study of hybrid diesel/solar/hydro/fuel cell energy schemes for a rural ICT Telecenter, *Energy*, **35**(2), pp. 639–646.
25. A. R. Prasad, E. Nataraajn (2006). Optimisation of integrated photovoltaic-wind power generation systems with battery storage, *Energy*, **31**(12), pp. 1943–1954.
26. Korea Electric Power Corporation (KEPCO) [Internet]. Available from: www.kepco.co.kr/eng/.
27. OPINET [Internet]. Available from: http://www.opinet.co.kr/index.do?cmd=main.

Chapter 2

Improved High-Efficiency Conventional Solar Photovoltaic System: A Novel Approach

Global environmental concerns and the escalating demand for energy, coupled with steady progress in renewable energy technologies, are opening up new opportunities for the utilization of renewable energy resources. Solar energy is the most abundant, inexhaustible, and clean of all the renewable energy resources till date. The amount of power from the sun intercepted by the earth is about 1.8×10^{11} MW, which is many times larger than the present rate of the entire energy consumption. Reducing prices of solar PV systems has been a constant challenge. Despite the recent advances, the prices of solar PV systems are still higher than those of conventional energy resources. In this chapter, a novel uninterruptible and environment-friendly wind-solar hybrid energy system (HES) with suitable cooling and reflector systems has been discussed. The solution was optimized by HOMER software using local solar and wind resources. The study presented in this chapter examines the effectiveness of three different structures/materials as reflectors: (i) silvered glass plane mirror, (ii) convex spherical

Solar Power and Energy Storage Systems
Hee-Je Kim

ISBN 978-981-4800-13-6 (Hardcover), 978-0-429-45877-4 (eBook)
www.panstanford.com

mirror, and (iii) aluminum (Al) foil. A comparative analysis of four different cooling techniques, that is, the water sprinkling system, the passive heat sink method, the active air fan method, and the closed-loop method, for enhancement of the output power was performed. A novel bireflector solar PV system (BRPVS) was suggested to control the working of the reflectors. In addition, the effect of using a reflector on the temperature of a solar PV system was studied. The combined enhancement of the output power by both the Al foil BRPVS and the cooling system was almost 22.8%–38.6%. An optimal control algorithm to use cooling and a BRPVS in an efficient manner is described.

2.1 Overall Survey

The use of solar photovoltaic (SPV) electric systems is growing rapidly in the sustainable renewable energy market, and these are expected to play an important role in the mix of future sustainable energy.[1,2] On the other hand, the initial capital cost of the photovoltaic (PV) modules and systems has always been a major barrier to the widespread use of this technology over the globe. The initial price of SPV systems can be reduced by producing more power out of the same PV module. The PV module's power output can be increased by increasing the incident solar radiation falling on a PV module according to the inherent characteristics of the PV cells. The distribution of solar radiation is not constant throughout a day. The sun changes its position from morning to evening, and its locus fluctuates throughout the year. A fixed installed solar panel with no sun tracking system cannot provide high output power. Single- and dual-axis solar trackers have been proposed for this purpose. On the other hand, installing a tracking system not only increases the initial cost of the SPV system as a number of motorized components are required but also reduces the life of the mechanical design of an SPV system frame because it will require a movable and adjustable structure. In addition to these drawbacks, the energy required for these motors to move the SPV system is also a problem. Power supply requires an energy storage system and a power converter system for smooth operation of these motorized tracking systems.

Previous studies reported that the intensity of solar radiation can be increased with a planar reflector system.[3] These reflectors will reflect light on the solar panel to enhance the electrical power output. The concentrating mirror booster technology is attracting increasing attention. The low cost and abundant availability make this option the most desirable. Nevertheless, inexpensive and cost-effective solutions are required to make installation more practical.

The initial focus of the solar thermal (ST) system was the output of a solar cell during winter.[4–6] Some researchers found that the alignment and placement of the solar panel at high latitudes with a vertical collector with a horizontal reflector.[7] Ray tracing models were also studied to calculate the increase in irradiation for a given reflector geometry.[7–10] Among these models, some apply diffuse reflectors that use a combined view factor and specular reflectance mode,[11,12] whereas some analyze the 2D specular reflectance model and some use experimental data.[7,10,13,14] The introduction of nonspecular corrugated booster reflectors enhances the results by up to 8%, and bidirectional reflectance function (BDRF) of these corrugated material is also noticed.[12] Enhancing radiation may improve the temperature of a PV module when the power and efficiency of the PV cell are decreased. Studies have shown that the efficiency of an amorphous silicon cell decreases by 0.05% with every 1°C temperature increase, and in the case of a crystalline silicon solar cell, the decrease ranges from 0.4% to 0.5%.[15] The effect of irradiance and temperature on output power (Fig. 2.1) is discussed by Wang et al.[16] A cooling system can enhance the performance of solar systems. Currently, the challenge for an SPV system is to design a cost-effective cooling system.

Different solutions have been proposed to decrease the operating temperature of a PV panel. Skoplaki et al.[15] calculated the dependence between the conversion efficiency and temperature of PV cells. A hybrid PV/thermal (PV/T) solar system is a reliable method for cooling PV panels.[17] In this system, the PV panels are attached to the cooling system. Water or air, which is used as a cooling agent, is circulated around the PV panel to cool the solar cells. The key factor of this system is that the warm water coming from around the solar panel can be used for domestic purposes. In addition, a previous study[18] proposed a cooling model in which water and air

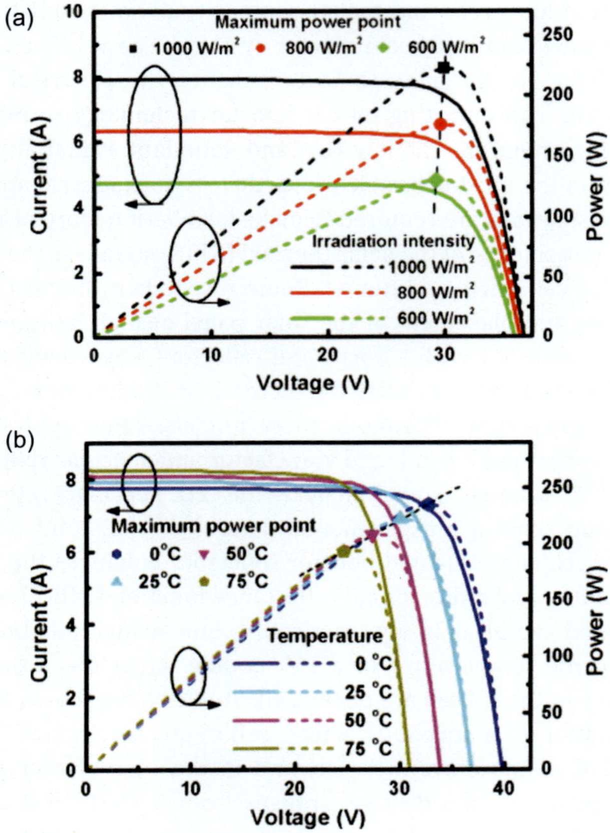

Figure 2.1 Effect of irradiance (irradiation intensity in W/m^2) and temperature on output power (W) of the PV module.[16]

were both observed as cooling agents. The performance of water as a cooling agent is higher than that of air as a cooling agent. The high performance of water is due to the increase in the glazing area of water. Similarly, air and passive cooling for improving the temperature of PV panels were assessed.[19]

Computational fluid dynamics was also used to examine the effects of airflow on the output of PV panels.[15] In this model, air flows under the panels to increase cooling by natural convention. Another system for improving panel efficiency was based on water as a coolant, in which two solar panel prototypes, one sprayed with water and the other without spray, were compared.[20] A microheat pipe array was designed for solar panel cooling,[18] which consisted of an evaporator and a condenser. The cool water in the evaporator section is vaporized by the input heat from the sun, which is then passed to a condenser section that is cooled using water or air. Therefore, the heat from the solar panel is transferred to air or water. The analysis showed that using water as a cooling agent decreases the temperature by 8°C and increases the efficiency by 3%. In the case of air, the temperature was decreased by 4.7°C and the efficiency was enhanced by 2.6%. Therefore, water is more effective as a cooling agent.[22]

Air cooling for PV panels was reported.[23] A ventilated channel, 0.1 m in width, behind each PV panel was used for cooling. Thus, a PV/T system was used, which is the most feasible solution for cooling PV panels, with the added advantage of producing thermal energy. This feature is more important because it generates two main sources of energy: electricity and heat. A bifacial silicon solar cell (Fig. 2.2) with an Al sheet reflector was studied by Moehlecke et al.[24]

Overall, there are different methods to cool SPV systems, and the use of reflectors can increase the output power of an SPV system considerably. Most cooling systems operate on some kind of water pumping system. The operation of this type of cooling system requires electrical power, which can be drawn from the SPV system. This reduces the effectiveness of the cooling system in increasing the SPV system efficiency. Furthermore, the cost of SPV systems increases. Continuous operation of this cooling system requires considerable electrical energy.

Energy software developed by the National Renewable Energy Laboratory, USA, called HOMER (Hybrid Optimization Model for Electric Renewables), was used to develop and examine the off-grid hybrid renewable energy system. There are several input parameters, such as electrical load demand, constituent demonstrated details, cost of all components related to the system, renewable energy resources as solar radiation data and wind speed data, generator specifications, battery specifications, and converter specifications, that have to be supplied to HOMER renewable energy software. An optimal pattern to provide the expected electrical

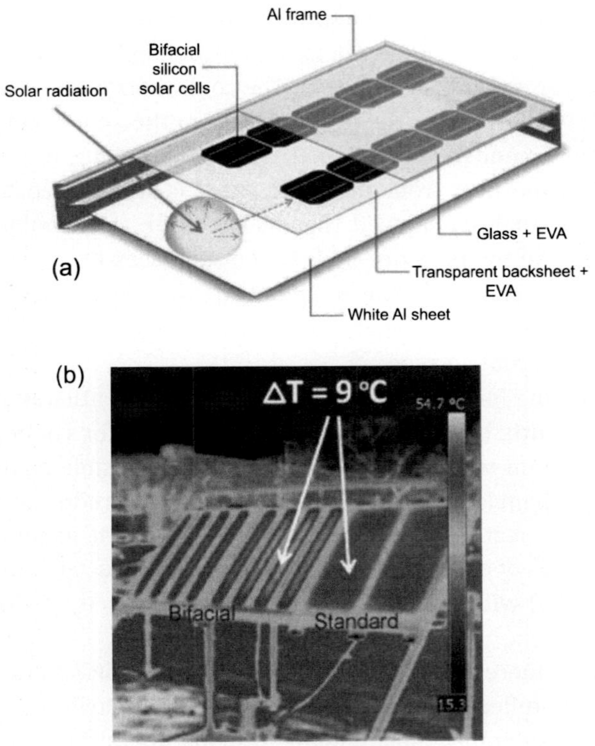

Figure 2.2 (a) Bifacial silicon solar cell with an Al sheet reflector; (b) thermographic images—a comparison with a standard PV module.[24]

load demand has been designed by the HOMER renewable energy software. HOMER executes a large number of hourly simulations to design a complete hybrid renewable energy system. To observe the impact of PV speculation cost, solar segregation, diesel fuel cost, and wind speed on the cost of energy (COE) HOMER has been executing sensitivity analysis by calculating lots of hourly data provided from different resources. To find an optimized solution, this study performed a number of simulations. The proposed solution was derived using wind and solar resources of the area and optimized using HOMER software with real-time experiments. This study performed real-time experiments and comparative analysis of different cooling methods proposed in the literature to integrate the solution with the proposed reflector materials, and its effects on the output power of an SPV system were examined. This study assessed three different solar reflector systems. The aim was to find an optimal material with the best design for future commercial amorphous and crystalline silicon SPV systems. In addition, this research assessed an optimal way to operate an SPV system with the proposed bireflector solar PV system (BRPVS) and cooling system. The heating rate model and BRPVS were based on the operating conditions, that is, solar radiation, ambient temperature, and SPV position.

The remainder of this chapter is organized as follows: An overview of the proposed reflector system and cooling techniques is provided in Section 2.2. Section 2.3 presents the proposed cooling systems. Section 2.4 discusses the hybrid energy system (HES). Sections 2.5, 2.6, and 2.7 discuss the experimental results and conclusions.

2.2 Bireflector Photovoltaic System

The use of planar reflectors is an already established technique for increasing the solar flux incident on a collector, and various studies have been performed for estimating the increase in irradiation from a given collector-reflector system geometry. Influence of the string mismatch losses, present at periods of heterogeneous illumination, on the yield of PV modules equipped with static planar reflectors was studied by Pavlov et al.[25] The modules under concentration represented here are situated in the middle of rows 2 and 3 and are, therefore, not subject to severe edge effects (see Fig. 2.3). Figure 2.4 presents the gain in the daily and monthly produced energy achieved by *p*-Si and *a*-Si PV modules under concentration, compared with the control modules.

Contrary to existing research, in our study three very inexpensive reflective materials were addressed for economical and high-output-power SPV systems. One material is a very thin sheet of aluminum, ranging from about 0.006 mm to the upper ISO defined limit of 0.2 mm (200 μm). To prevent oxidation, the aluminum oxide layer is added on top of Al during the manufacturing process of Al foils. Aluminum foil is produced from cast ingots that are rolled to the

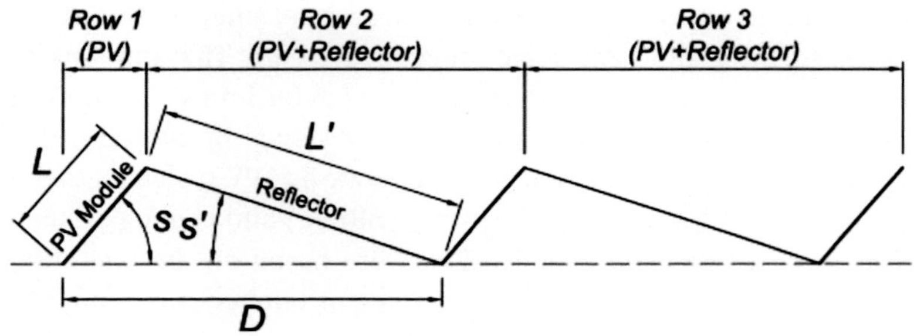

Figure 2.3 Schematic of the experiment. PV module rows 2 and 3 are equipped with reflectors; row 1 serves as the control.[25]

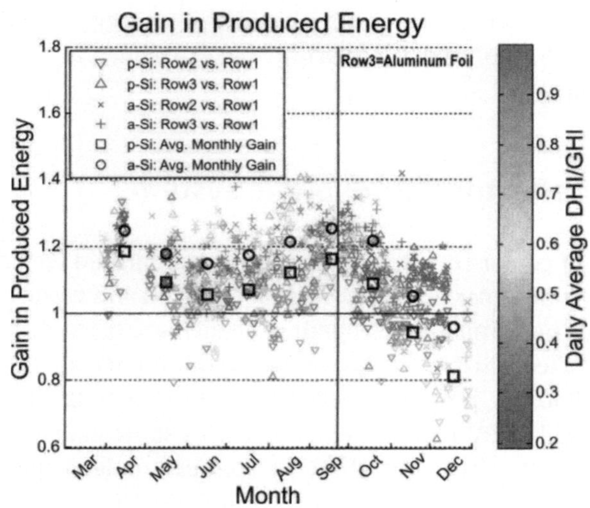

Figure 2.4 Experimental data: gain in daily and monthly produced energy resulting from concentration. The color scale represents the daily ratio of diffuse irradiation to global irradiation.[25]

required thickness using a beta radiation sensor. According to the radiation intensity, it rolls the sheet and the roller adjusts the thickness. Silvered glass mirror is the second material to be used as a reflector to enhance the output power of the solar panel. From all available shapes, a plane and convex spherical silvered glass mirror was used. Silver nitrate was used to manufacture the mirror. Silver-coated glass mirror has many properties that help to improve the

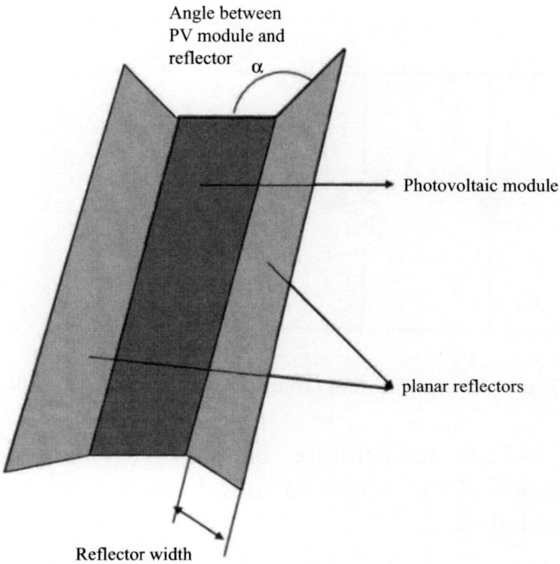

Figure 2.5 An overview of a bireflector solar PV arrangement.

efficiency, such as ease of fabrication, rigidity, hardness, and ability to take a smooth finish. For more protection, the reflective coating was applied to the back of the glass. A beam of light reflects off a mirror at an angle of reflection equal to its angle of incidence, that is, if the beam of light is shining on a mirror surface at a vertical angle, it reflects from the point of incidence at an angle in the opposite direction.[26,27] A convex mirror helps scatter the beam of light. Figure 2.5 presents an overview of the reflectors attached with the SPV module discussed in this chapter.

2.2.1 Electrical Modeling

A BRPVS can be modeled by its equivalent electric circuit. Solar cells are replaced by a current source with a photogenerated current I_L with a p-n junction diode connected in parallel. This also has a parallel resistance, R_{sh}, and a series resistance, R_s, as shown in Fig. 2.6.

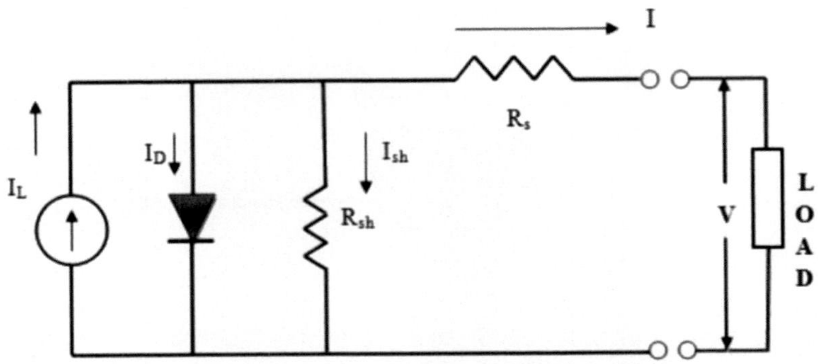

Figure 2.6 Equivalent circuit for an individual PV cell.

At a constant temperature and solar irradiance, the V–I characteristics of the proposed model can be given using the following relation:

$$I = I_L - I_o \left[\exp\left(\frac{V + R_s}{a}\right) - 1\right] - \frac{V + R_s}{R_{sh}} \tag{2.1}$$

The resistances represent the external behavior of the model. Values of important parameters ($I_{L,ref}$, $I_{o,ref}$, $R_{s,ref}$, $R_{sh,ref}$ and a_{ref}) for modeling a PV module with a BRPVS can be estimated by using conditions in Table 2.1.[28]

On applying all five conditions in Eq. 2.1, the following equations are obtained [3]:

$$I_{sc,ref} = I_{L,ref} - I_{o,ref} \left[\exp\left(\frac{I_{sc,ref} R_{s,ref}}{a_{ref}}\right) - 1\right] - \frac{I_{sc,ref} R_{s,ref}}{R_{sh,ref}} \tag{2.2}$$

$$I_{L,ref} = I_{o,ref} \left[\exp\left(\frac{V_{oc,ref}}{a_{ref}}\right) -1\right] - \frac{V_{oc,ref}}{R_{sh,ref}} \tag{2.3}$$

Table 2.1 Conditions used to estimate parameters

At short circuit	$[dI/dV]_{sc} = -1/R_{sh,ref}$
At open circuit voltage	$I = 0, V = V_{oc,ref}$
At short circuit current	$I = I_{sc,ref}, V = 0$
At the mpp	$I = I_{mp,ref}, V = V_{mp,ref}$
At the mpp	$[dI/dV]_{sc} = 0$

$$I_{\mathrm{m,ref}} = I_{\mathrm{L,ref}} - I_{\mathrm{o,ref}}\left[\exp\left(\frac{V_{\mathrm{mp,ref}} + I_{\mathrm{mp,ref}}R_{\mathrm{s,ref}}}{a_{\mathrm{ref}}}\right) - 1\right] - \frac{V_{\mathrm{mp,ref}} + I_{\mathrm{mp,ref}}R_{\mathrm{s,ref}}}{R_{\mathrm{sh,ref}}} \quad (2.4)$$

$$\left[\frac{dI}{dV}\right]_{\mathrm{sc}} \cong -\frac{1}{R_{\mathrm{sh,ref}}} \quad (2.5)$$

$$\frac{I_{\mathrm{m,ref}}}{V_{\mathrm{mp,ref}}} = \frac{\left(\frac{I_{\mathrm{o,ref}}}{a_{\mathrm{ref}}}\right)\exp\left(\frac{V_{\mathrm{mp,ref}}+I_{\mathrm{mp,ref}}R_{\mathrm{s,ref}}}{a_{\mathrm{ref}}}\right) + \frac{1}{R_{\mathrm{sh,ref}}}}{1 + \left(\frac{I_{\mathrm{o,ref}}R_{\mathrm{s,ref}}}{a_{\mathrm{ref}}}\right)\exp\left(\frac{V_{\mathrm{mp,ref}}+I_{\mathrm{mp,ref}}R_{\mathrm{s,ref}}}{a_{\mathrm{ref}}}\right) + \frac{R_{\mathrm{s,ref}}}{R_{\mathrm{sh,ref}}}} \quad (2.6)$$

The maximum power output (MPP) can be obtained from MPP current (I_{mp}) and voltage (V_{mp}) using the following relation:

$$P_{\mathrm{mp}} = I_{\mathrm{mp}}V_{\mathrm{mp}} \quad (2.7)$$

In estimating the PV module performance, the temperature dependence of the maximum power point efficiency is an important parameter and is given by the following:

$$\eta_{\mathrm{mp}} = \frac{I_{\mathrm{mp}}V_{\mathrm{mp}}}{G_{\mathrm{T}}A_{\mathrm{m}}} \quad (2.8)$$

2.2.2 Optical Modeling

The proposed BRPVS can be modeled easily by considering the conventional optical geometry approach. The total solar radiation (S_{tot}) absorbed by the BRPVS is equal to the sum of the direct radiation (S_{d}) on the PV surface, the ground reflected radiation (S_{g}), the sky-diffuse radiation (S_{SK}), the reflected radiation from the right reflector (S_{reflR}) to the surface of the PV panel with a tilted plane angle (α_1), and the radiation reflected from the left reflector (S_{reflL}) to the PV panel with a tilted plane angle α_2:[28]

$$S_{\mathrm{tot}} = S_{\mathrm{d}} + S_{\mathrm{g}} + S_{\mathrm{sk}} + S_{\mathrm{reflR}} + S_{\mathrm{reflL}} \quad (2.9)$$

A previous study[29] discussed the BRDF for the optical modeling of an SPV system with a reflector. Schlick[30] presents an integrative method in the analysis, where the contribution of irradiation to a differential point on the PV module (dx) from each differential scattering element on the reflector (dy) in the direction of dx is calculated. Therefore, to determine the total solar irradiations on the surface of the module, integration was performed along the two principal directions of the array, x and y. Equation 2.10 shows the energy incident on the surface of the module (E_r^{i}) that is impacting the plane reflector at point dx.

$$E_r^{\mathrm{i}} = E_{\mathrm{p}} \cos(\theta')\mathrm{d}y\mathrm{d}z \tag{2.10}$$

Equations 2.11 and 2.12 show the value of the reflected radiant intensity per unit depth through a differential angle (E_{ref}) and the radiant intensity per unit depth that strikes the surface of the module (E_{m}), respectively.[31]

$$E_{\mathrm{ref}} = E_{\mathrm{r}}^{i}\mathrm{BDRF}(\theta, \theta')\rho \tag{2.11}$$

$$E_{\mathrm{m}} = E_{\mathrm{ref}}\mathrm{d}\theta \tag{2.12}$$

Equation 2.14 is defined in terms of the angle of incidence (θ) of the differential ray onto the surface of the module and the distance the ray has traveled.[31]

$$\mathrm{d}\theta = \frac{\mathrm{d}y \cos(\beta)}{r} \tag{2.13}$$

$$\mathrm{BDRF}\left(\theta, \theta'\right) = \frac{D(\theta, \theta')}{\int_{-\frac{\pi}{2}}^{\frac{\pi}{2}} D(\theta, \theta')\mathrm{d}\theta} \tag{2.14}$$

Equations 2.10 to 2.14 can be combined and integrated along the characteristic dimensions and multiplied by the depth of the module being analyzed (A_{module}). Irradiance (W/m^2) on the surface of the module due to reflection (G_m^i) can be calculated using the following relation[32]:

$$G_m^i = \frac{G_{\mathrm{p}}\rho}{A_{\mathrm{m}}} \int_0^{L_{\mathrm{sect}}} \int_{L_{\mathrm{m,min}}}^{L_{\mathrm{m,max}}} \mathrm{BDRF}(\theta, \theta') \cos(\theta') \frac{\cos(\beta)}{r} \mathrm{d}x\mathrm{d}y \tag{2.15}$$

Here A_{m} is the area of the module, ρ is the spectacularly constant surface reflectivity, and θ' is the angle between incoming ray and reflector surface.

2.3 Proposed Cooling Techniques

The performance of PV modules is generally specified under standard test conditions (STC); however, the performance of the modules under real field conditions differs from the expectations derived from the results under STC due to a variety of continuously changing conditions. Therefore, it is important to monitor the behavior of PV modules under real field conditions with temperatures higher than the ambient temperature defined at STC (25°C). An increase in temperature of the PV module causes the bandgap of the intrinsic semiconductor to shrink, and as a result the open circuit voltage (V_{oc}) decreases. At the same time, a lower bandgap allows more incident energy to be absorbed because a larger percentage of incident light has sufficient energy to raise charge carriers from the valence band to the conduction band; a larger photocurrent results. As the temperature is raised, however, the internal resistance of the material increases and the electrical conductivity decreases. The increase in current for a given temperature rise is proportionally lower than the decrease in voltage. Hence, the efficiency of the cell is reduced. Cooling systems are required to reduce the temperature and increase the efficiency of an SPV module. On the basis of a literature survey and previous work, the following four cooling techniques were compared in an effort to achieve the most effective cooling system for an SPV system.[33]

2.3.1 Passive-Air-Cooled System

A passive-air-cooled system for solar panels involves the installation of heat sinks at the rear side of the solar panel. A heat sink transfers thermal energy from the higher-temperature rear side of a solar panel to the lower-temperature surrounding air. For the effective utilization of heat sinks, the solar panel temperature must be higher than the that of the surroundings to transfer heat by convection, radiation, and conduction. Cooling a solar panel by using a heat sink is getting attention these days. Various locations on the surface were considered for spectral distribution measurement (spectrum in Fig. 2.7a) and irradiance distribution (bar chart in Fig. 2.7b). A halogen spectrum provides information about the intensities and

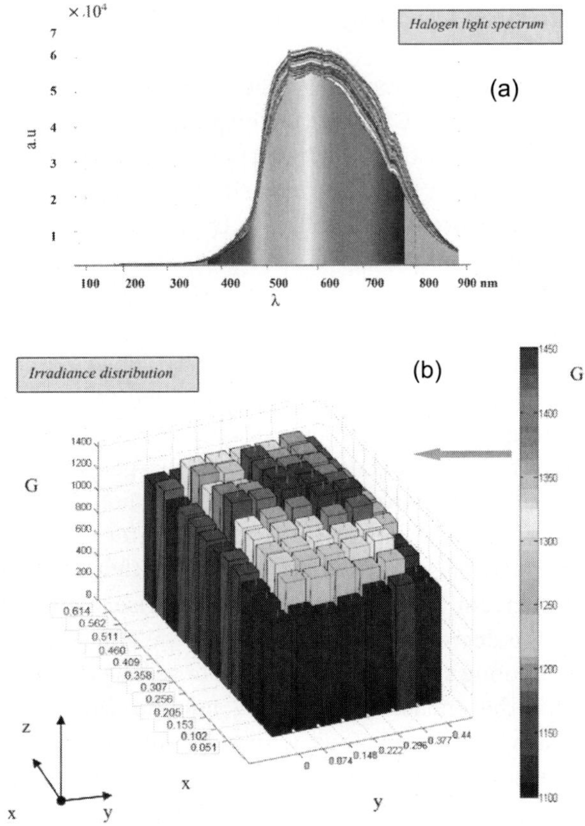

Figure 2.7 (a) The halogen light spectral distribution and (b) irradiance distribution on the devices.[34]

nonuniformity of ultraviolet, visible light, and near-infrared (IR) radiation from a basic data reduction.[34]

The geometry of the model heat sink–based SPV cooling is investigated by Popovici et al.[35] Figure 2.8a was realized using ANSYS – Design Modeler. The ribs of the heat sink have circular holes of 0.003 m radius, placed at a distance of 0.03 m from each other. These holes intend to improve the air circulation near the heat sink and to extract more heat from the PV panel. The mesh was realized with different refinement for the heat sink and ventilation channel. Therefore, the minimum size of the cell in the domain of interest

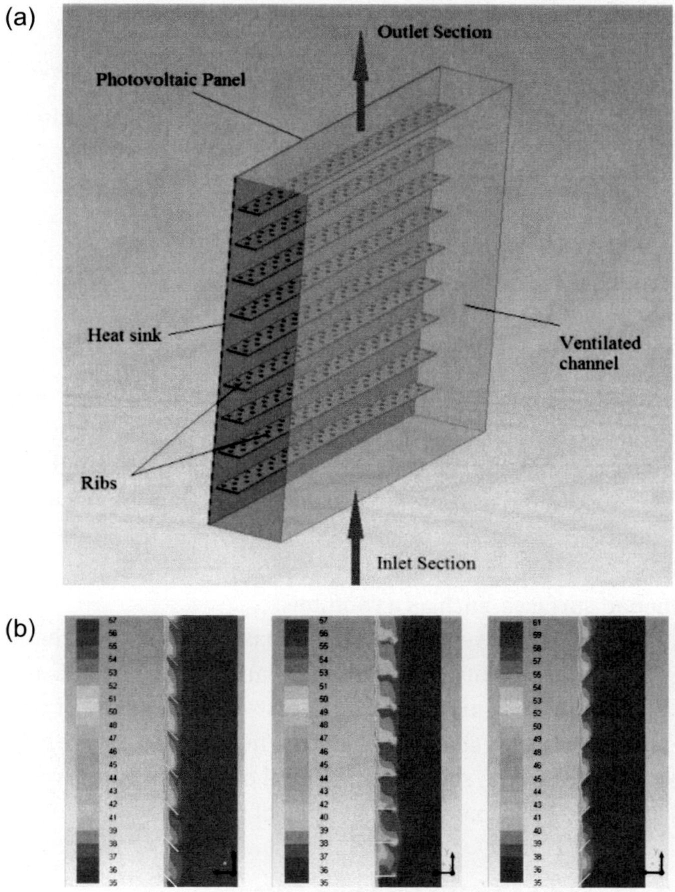

Figure 2.8 (a) Geometry of the studied model[35] and (b) temperature spectra for different angles of ribs: (a) 45°, (b) 90°, and (c) 135°.[35]

is about 0.002 m for the heat sink and ribs and about 0.008 m for the air channel. Results of simulations consist of the PV panel temperature variation depending on the height and angle of the ribs (see Fig. 2.8b).

Heat transferred by a heat sink can be explained by Fourier's law of heat conduction. The rate at which heat is transferred by conduction (q_k) is proportional to the product of the temperature gradient ($\frac{\mathrm{d}T}{\mathrm{d}x}$) and the cross-sectional area through which heat is transferred.

$$q_k = -kA\frac{dT}{dx} \tag{2.16}$$

Applying the conservation of energy, for steady-state conditions, and Newton's law of cooling to the temperature node of the heat sink in a duct gives the following two equations:

$$\dot{Q} = \dot{m}\, c_{p,in}(T_{air,out} - T_{air,in}) \tag{2.17}$$

$$\dot{Q} = \frac{T_{sh} - T_{air,av}}{R_{sh}} \tag{2.18}$$

Here $\dot{m}$ is the air mass flow rate in kg/s.

The fin efficiency is one of the parameters that make a higher-thermal-conductivity material important. The fin efficiency increases with a decreasing fin aspect ratio (making them thicker or shorter) or on use of a more conductive material (copper instead of aluminum). Many shapes exist in the literature, ranging from elliptical and cylindrical cylinders and conical shapes to more sharpened surfaces, such as a rhombus.

Square sections were used with a base thickness in the 5 mm range and a fin height of 45 mm, with an air spacing of 5 mm. To utilize the heat sink, an optimal and cost-effective array of the heat sink was arranged as shown in Fig. 2.8. The heat sinks were attached to the rear side of the solar panel. A total of 26 heat sinks were used for a 150-watt solar panel with an area of 0.991 m^2.

2.3.2 Closed-Loop Water Loop System

In our experimental study we tried to combine conventional PV and thermal heating methods to develop a very low cost multiobject (PV cooling and water heated) system as conventional PV electricity generation and solar heating are still very expensive. A lot of investment is required for developing ST and SPV. For instance, Fig. 2.9 shows a comparison of the investment costs for an ST system and a solar electric system PV in Freiburg, Germany.[36] To achieve this goal of effective SPV cooling and water heating the PV module was integrated with a pipe (for cooled water circulation) with a thickness of approximately 20 mm attached to the rear side of the module. The water to be supplied to the cooling panel was stored in a storage tank

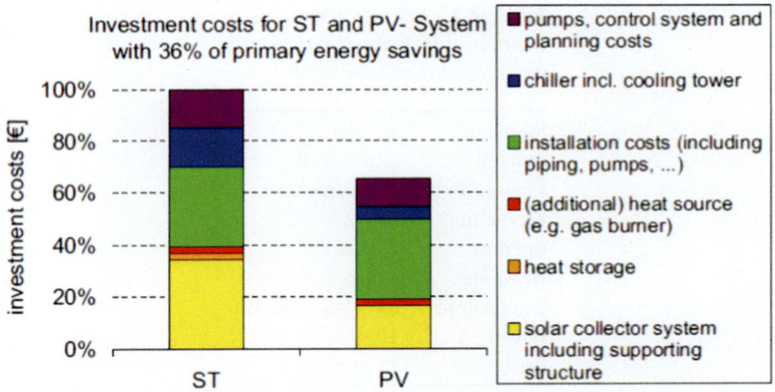

Figure 2.9 The investment cost for a solar thermal system and a solar electric system PV to achieve primary energy saving of 36% in Freiburg, Germany.[36]

with an insulation thickness of 10 mm. The outlet of the tank was connected to a pump for circulating water at the required pressure. To regulate the water flow inside the cooling panel, a flow meter with a maximum flow rate of 8.33 L/min. was used. The cooling water circulated through the collector, captured the waste heat from the PV module, and produced hot water collected at the collector outlet. An overview of the cooled-loop water-cooled system is given in Fig. 2.9.

The temperatures at various points of the system were measured using a submersible digital thermometer (mode no. Tp101). A small aquarium waterproof pump was used. Table 2.2 lists the specifications of the pump used for this setup.

2.3.3 Active-Air-Cooled System

Some researchers examined the effectiveness of an air-cooled system for a solar panel cooling system. The air cooling mechanisms in indoor testing used halogen lamps as a light source. The impact of the air cooling mechanism was analyzed under different levels of solar radiation and surrounding temperature. Using a brushless DC fan to reduce the PV panel temperature, the power output was observed with increasing solar radiation.[37]

Table 2.2 Specifications of a submersible waterproof pump used for closed-loop cooling

Model no.	SL-389
Voltage	220–240 V
Flow rate	500 L/h
Frequency	50 Hz
Current	0.07 A
Head max	1.6 m
Dimensions	64 × 45 × 56 mm
Power	10 W
Outlet	13 mm dia

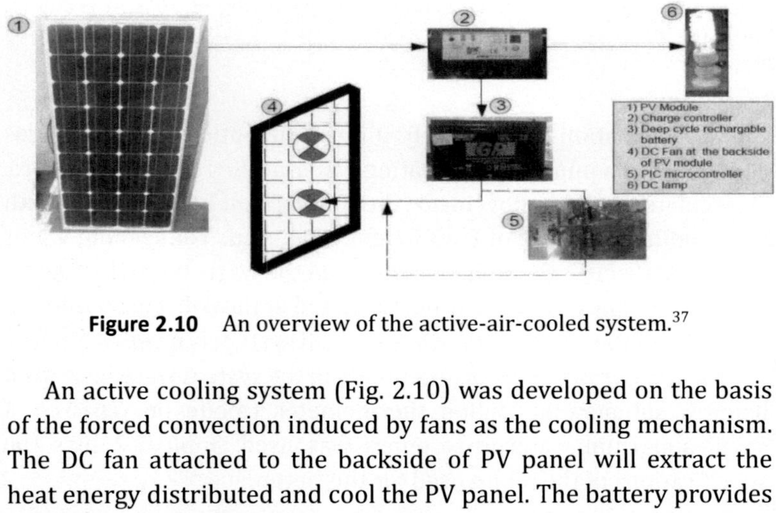

Figure 2.10 An overview of the active-air-cooled system.[37]

An active cooling system (Fig. 2.10) was developed on the basis of the forced convection induced by fans as the cooling mechanism. The DC fan attached to the backside of PV panel will extract the heat energy distributed and cool the PV panel. The battery provides the DC source to the forced active-air cooling system installed at the backside (DC brushless fan) of the PV module. The DC cooling system reduces the temperature to improve the output power of the PV module. Two different approaches for fan cooling systems were constructed using eight fans with the specifications shown in Table 2.3.

The DC brushless fans were arranged in such a way that a smaller number of fans could provide air to the entire panel. The fans were inverted to act like ordinary room fans to provide cooled air to the panel from the outside atmosphere. The increasing fan

Table 2.3 Specifications of a submersible waterproof pump used for closed-loop cooling

Dimensions (l × h × w)	120 × 120 × 25 mm
Voltage (V)	12 V DC
Safety current (A)	0.16 A + 10%
Rated current (A)	0.15 A + 10%
Power (watts)	1.92 W + 10%
Speed (RPM)	1200 RPM + 10%
Air flow (CFM)	44.73
Air pressure (mmH_2O)	1.65
Noise (dB)	19.1
Weight (g)	120
Bearing type	Sleeve
Life (hours)	30,000 h

capacity increases the overall heat transfer and PV panel cooling performance, but it is not advisable to increase the number of fans as it will increase the power consumption from the battery and reduce the overall efficiency of the system.[26]

2.3.4 Water Sprinkling System

Cooling the SPV system with water is the most discussed cooling technique. The percentage decrease in temperature of a solar panel depends on the cooling rate of the SPV material. Indoor test of the water sprinkling method has already been performed with the experimental setup as shown in Fig. 2.11a,b.[38]

The cooling period can be determined by obtaining precise information of the cooling rate of the SPV system. The energy balance equation provides a base to determine the cooling period. The amount of heat absorbed by cooling water (Q_{ab}) is equal to the amount of heat dissipated by the SPV system (Q_d). The cooling period can be determined using the following set of equations:[26]

$$Q_{ab} = Q_d \tag{2.19}$$

$$\dot{m} \times t \times C_w \times \Delta T_w = m_g \times C_g \times \Delta T_g \tag{2.20}$$

Here $\dot{m}$ is the mass flow rate of water, ΔT_w is the change in water temperature, m_g is the mass of glass, C_g is the heat capacity of

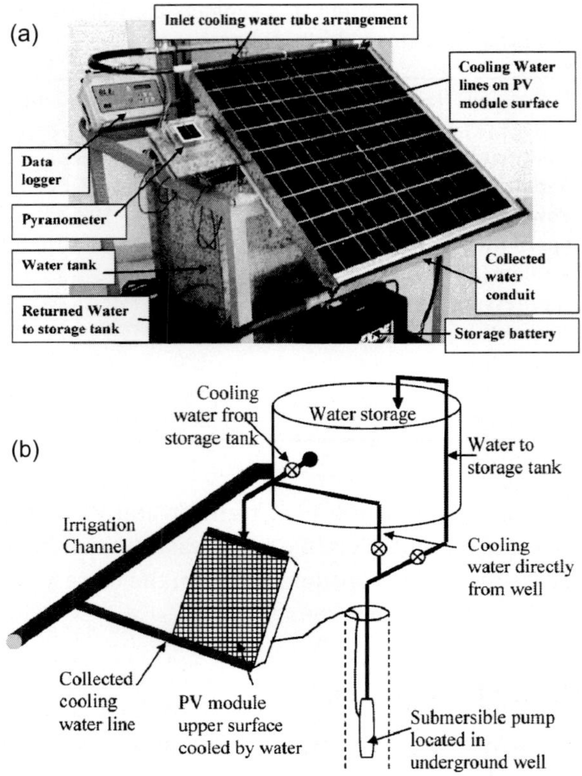

Figure 2.11 (a) PV water cooling test rig; (b) proposed PV-powered water pumping system configuration.[38]

glass, C_w is the specific heat capacity of water, ΔT_g is the change in glass temperature, and t is the time taken to adjust the solar panel temperature to a specified temperature.

Solar panels are manufactured from silicon, with the top covered completely by glass so that the physical properties of glass are taken as the physical properties of the solar panels.

2.4 Hybrid Energy Systems

A number of authors have used HOMER for optimizing hybrid systems for different locations worldwide. HOMER is found to be

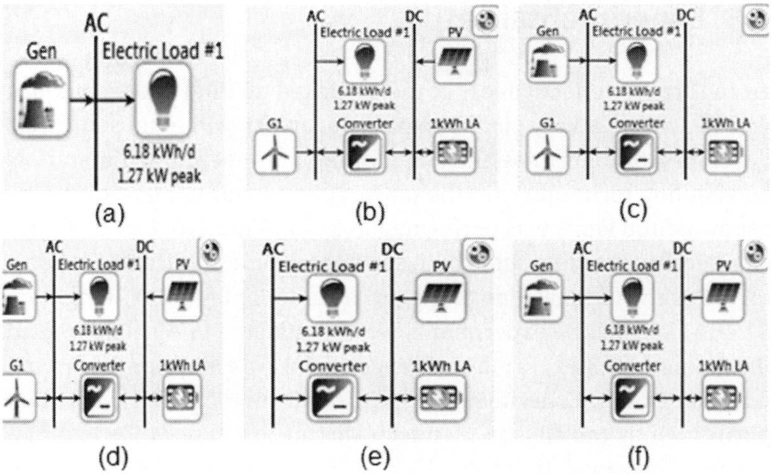

Figure 2.12 Schematic diagram of the optimized power systems simulated in HOMER software: (a) base/reference system, (b) solar-wind hybrid (SPV-W), (c) wind generator (W-GEN), (d) solar-wind generator (SPV-W-GEN), (e) solar (SPV), and (f) solar generator (SPV-GEN).

one of the most widely used software for the optimization and sensitivity analysis of hybrid systems. The inputs required for the analysis are solar radiation, wind speed, temperature data, load profile, constraints, system control, and economic factors. The HOMER software/database was used to search for optimized solutions based on the available resources. On the basis of the optimized system suggested after a brief analysis of local renewable resources by HOMER, five different configurations were considered as comparative and optimized solutions, as depicted in Fig. 2.12: (i) solar-wind-generator (SPV-W-GEN) HES, (ii) SPV system, (iii) solar-wind hybrid (SPV-W), (iv) solar-generator (SPV-GEN) HES, and (v) wind generator (W-GEN). All five systems were off-grid, and the size of the system was optimized in such a way that all the energy generated by the systems used served the electrical load of houses where these systems were installed to avoid any expenses related to grid expansion. A comparative analysis was performed by comparing the outcomes of these five systems, with the base system comprising a generator serving the load.

2.5 Experimental Setup

To confirm the effectiveness of the proposed method and techniques and to achieve a very clear comparison, an experimental setup was developed to determine the effect of the aforementioned materials and cooling techniques on the output power and temperature. The experimental work was performed in front of the 107 mechatronics engineering building and the laser sensor and application laboratory, Pusan National University, at latitude 31.04083°N, longitude 31.4861303°E. The experiments were conducted in April 2016, and the azimuth of 304° (a) in each experiment was measured from the north in Busan. Experiments with a slight modification were then performed at the Nelson Mandela African Institute of Technology, Arusha, Tanzania, (latitude 3°23′58″S, longitude 36°47′48″E), in January 2017. The tilt angle varies with location, and optimal tilt angles of 30° and 0° were selected for the experiment on a solar panel facing south 0° at Busan and Arusha, respectively. Initially, a commercial PV module with a capacity of 20 watts was used to determine the effect of the reflectors on its performance at the given latitude. The electrical specifications of the module used for analysis of the bireflector system are given in Table 2.4 under STC.

A MASTECH MS8217 millimeter was used to measure the voltage and current under various irradiance and reflector conditions. A FLUKE VT04A visual IR thermometer was used to record the thermal images at a specified time. Three different materials and structures were studied. The voltage and operating temperature of the PV module were recorded and saved using an EL-USB-3 voltage data logger and EL-USB-3 temperature and humidity data logger.

Table 2.4 Specifications of a solar panel

Short circuit current	1.38 A
At open circuit voltage	21.5 V
Maximum power voltage	17.5 W
Maximum power current	1.15 A
Maximum power	20 W
Test conditions AM 1.5 25°C	1000 W/m^2

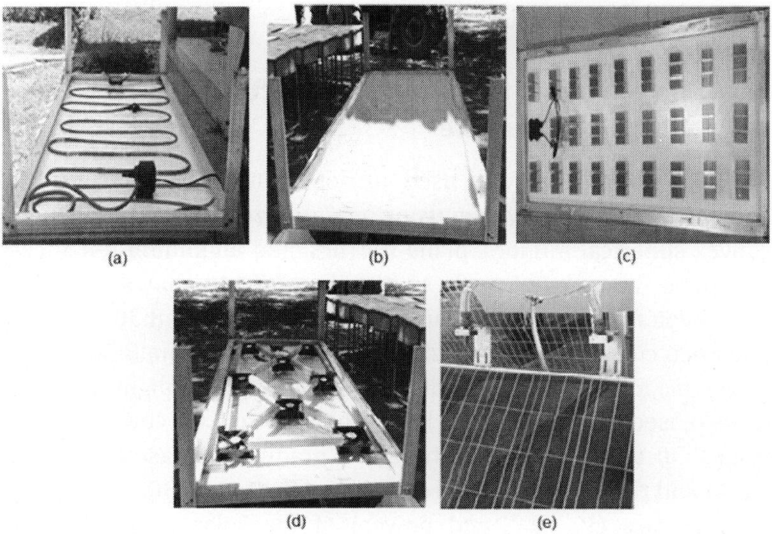

Figure 2.13 Experimental setup for a comparative analysis of proposed cooling systems, that is, (a) a closed-loop system, (b) no cooling, (c) heat sink–based passive cooling, (d) an active-air-cooled fan system, and (e) a watering sprinkling system.

An aluminum foil sheet, 0.5 mm in thickness, was used to make the structure firm. The foil was wrapped around a rectangular hard paperboard. For the very first experiment, the area (LXW) of the aluminum foil was kept the same as that of the solar panel (0.54 m × 0.36 m). A silver-coated plane mirror of 0.1944 m^2 area was used. The convex mirror scatters light, so two convex mirrors, 15 cm in diameter each, were fixed together and used as a reflector, as shown in the figure. For the real-time installation of SPV systems, perfect south (180°N) is not available, so three different positions were studied: southwest 210°N (30° diverted toward the west) and southeast 150°N (30° diverted from the south toward the east). A VICTOR 816 digital anemometer was used to measure the wind velocity of the surrounding air. Detailed solar insolation and weather data were obtained from a weather station installed. The experimental setup acquired data continuously from sunrise to sunset.

2.6 Results and Discussions

2.6.1 Investigating the Effective Reflector Material/Structure

A 20 W solar panel was used to determine the effects on the outcomes of the SPV system using an optimized BRPVS made up of convex spherical mirrors, plane mirrors, and aluminum foil sheet. The panel was placed at 0° south. Figure 2.14a shows the open circuit voltage, short circuit current, and power from 8:30 to 16:30. The open circuit voltage remains almost constant from 8:30 a.m. to 4:00 p.m., followed by a sudden decrease as the sunlight intensity is decreased at its maximum level. The short circuit current varies from 8:30 to 12:00, where there is a gradual increase (0.67 A to 1.2 A) and then a gradual decrease from 12:00 to 16:30.

Outcomes (V_{oc}, I_{sh}, and power) of the solar panel with no reflector were then compared with outcomes of solar panels with plane, silvered mirrors, Al foil sheets, and convex spherical mirrors as reflectors (see Fig. 2.14). The open circuit voltage (V_{oc}) and short circuit current, I_{sh}, for the plane, spherical mirror and the Al foil reflector were calculated from 8:00 a.m. to 4:30 p.m. A significant enhancement in the output power using the reflector was noticed. As three different types of reflectors were used, the effects of each reflector material and structure were different. The maximum power delivered by the solar panel with no reflector was 16.84 W. The maximum power delivered by the convex spherical mirror, the silvered-coated plane glass, and the Al foil reflector was 17.44 W, 18.91 W, and 20.31 W, respectively. The spherical mirror has much less of an effect, and the Al foil showed a comparatively larger effect than the plane mirror. The short circuit current and power curve showed this effect. Moreover, the reflector systems are more effective in the morning and evening and comparative less effective at noon.

Experimental studies showed that use of a reflector is inexpensive and effective in enhancing the output power of an SPV system. On the other hand, the effect of these reflectors on the solar panel temperature is unclear. Real-time IR thermal images of each reflector system were collected using the maximum ambient

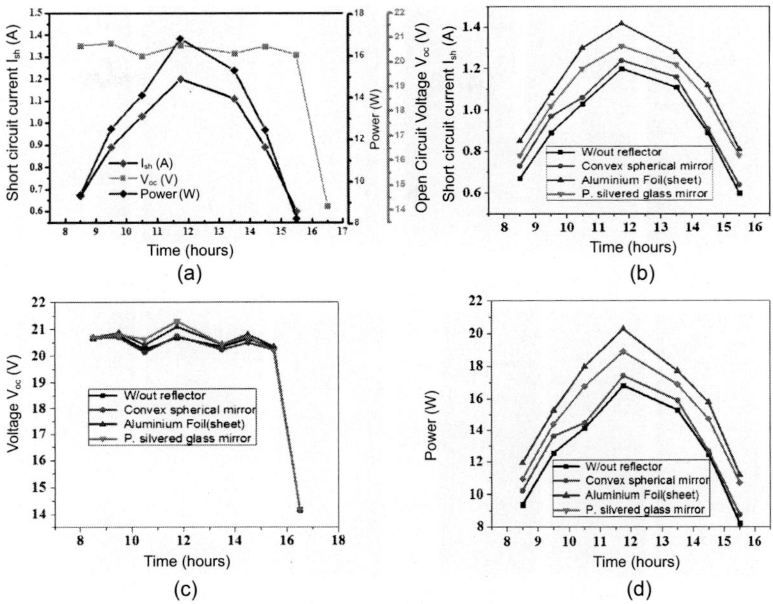

Figure 2.14 (a) Solar panel open circuit voltage (V_{oc}), short circuit current (I_{sh}), and output power without any reflector; comparative analysis of the (b) short circuit current (I_{sh}); (c) open circuit voltage (V_{oc}); (d) output power of silvered glass plane mirror, convex spherical mirror, and aluminum foil.

temperature. The peak temperature time for investigating these figures was selected.

Thermal images (Fig. 2.15) at 11:30 showed the maximum hotspot temperature for a solar panel without a reflector, a spherical mirror, a plane mirror, and an Al foil sheet of 42.1°C, 42.3°C, 42.9°C, and 43°C, respectively. At 14:37, the maximum hotspot temperature for a solar panel without a reflector, a spherical mirror, a plane mirror, and an Al foil sheet was 38.9°C, 38.9°C, 39.4°C, and 40.5°C, respectively. Similarly, at 16:35, the maximum hotspot temperature for the solar panel without a reflector, a spherical mirror, an Al foil sheet, and a plane mirror was 27.1°C, 27.2°C, 27.3°C, and 27.5°C, respectively. These images show that the use of a reflector does not have a large effect on the temperature of the solar panel and the use of reflectors is safe for SPV operation.

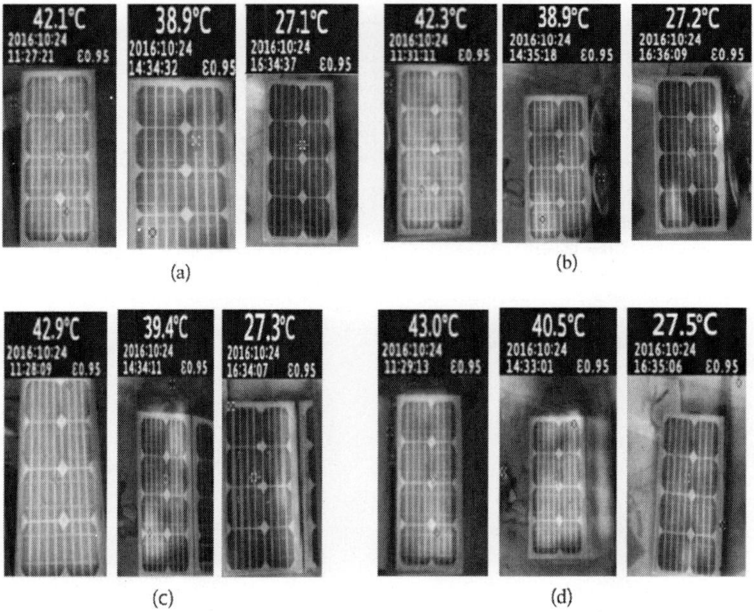

Figure 2.15 Effect of a reflector on the temperature of a solar panel. (a) Without reflector; (b) convex spherical mirror; (c) silvered glass plane mirror; (d) Al foil sheet.

2.6.2 Investigating the Appropriate Cooling Technique

The proposed cooling system was analyzed using real-time experiments, and the outcomes were compared. The experiments were performed at Arusha, Tanzania. All four cooling systems were tested on the same day and time to obtain accurate data for the open circuit voltage and short circuit current. The temperatures of the solar panels with different cooling systems were recorded using an EL-USB-3 (temperature and humidity data logger), and thermal images were taken using a FLUKE VT04A visual IR thermometer. The operation of the passive heat sink method does not require electric power to cool the SPV module. Initially, the data from the SPV system with no cooling or natural cooling and the data (SPV module) with a specifically arranged heat sink were monitored from 8:00 to 10:30 a.m. at specific time intervals while the other three systems were turned on when the temperature of the panel reached

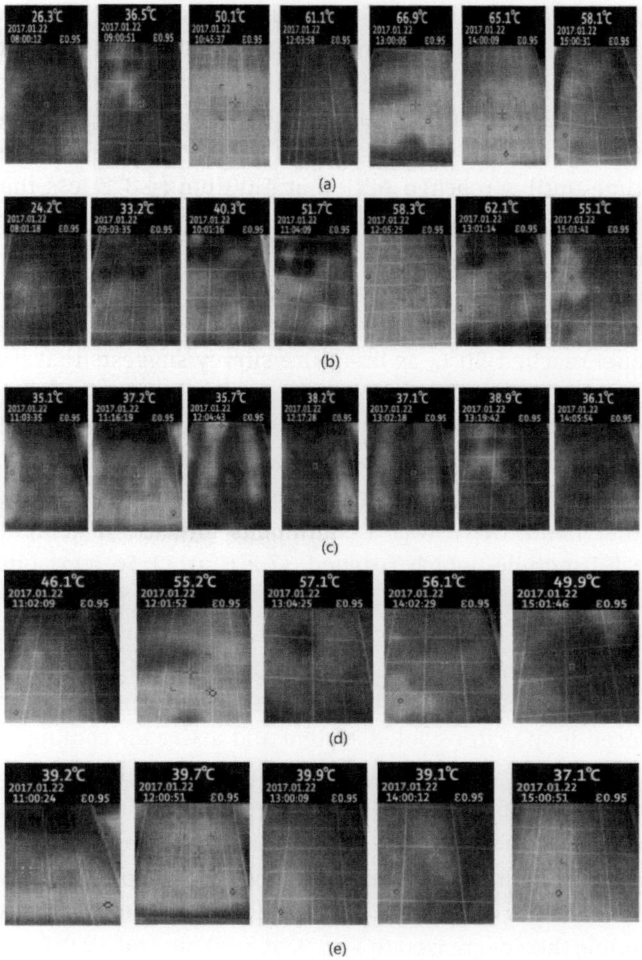

Figure 2.16 Infrared thermal images of different cooling systems: (a) no cooling, (b) passive heat sink method, (c) water sprinkling system, (d) closed-loop system, and (e) active air fan cooling system.

50°C. The thermal images for the PV module in Fig. 2.16a showed that the temperature of the PV module was 26.3°C at 8:00 am, which increased with time until it reached 66.9°C at 1:00 p.m. At 1:00 p.m., the temperature of the PV module was 41.9°C more than that under STC (25°C). The temperature of the module then decreased to 65.1°C at 2:00 p.m. and 58.1°C at 3 p.m. The passive heat sink method reduces the temperature to some extent, as shown in Fig. 2.16b.

The temperature of the PV module cooled by the heat sink method was 24.2°C at 8:00 a.m. (2.1°C less than the panel with no cooling system). The temperature of the module kept increasing with time until it reached 62.1°C at 1:00 pm (4.8°C less than the panel with no cooling). At 1:00 p.m., the temperature of the PV module was still 37.1°C more than that under STC (25°C). The temperature of the module then decreased to 55.1°C at 3:00 p.m. The cooling systems were not turned on before the temperature of the panel reached 50°C, as literature survey suggests that it is not economical to use a cooling system at a low temperature.[29]

The temperature of the PV module reached 50.1°C at 10:45 a.m. The temperature of the PV module was lowered to 35.1°C (11:03 a.m.) by the water sprinkling method for cooling (Fig. 2.16c). On the other hand, there was a continuous increase in temperature of the PV module, which reached 37.2°C at 11:16 a.m., until the water was sprinkled on the module at 11:30 a.m. The temperature of the PV module was kept in the 35.1°C–39.8°C range by sprinkling water every 30 min. Figure 2.16 compares the SPV modules with different cooling systems. Compared to the water sprinkling method, the closed-loop water-cooled system and active air cooling by fans require continuous operation. The temperature of the PV module was lowered to 46.1°C at 11:02 a.m. by the closed-loop method. The temperature of the module kept increasing with time until it reached 57.1°C at 1:00 p.m. At 1:00 p.m., the temperature of the PV module was 32.1°C more than that under STC (25°C). The temperature of the module then decreased to 56.1°C at 2:00 pm and 49.9°C at 3 p.m. Cooling the solar panels by air reduced the temperature of the panel to 39.2°C, and this temperature was maintained for the solar panel. Although a slight variation occurs, it still keeps the temperature in the 39.2°C–40°C range. The temperature data show that the water sprinkling method and the air-cooled method are more effective than the heat sink and the closed-loop methods for cooling the SPV system. Between cooling by water sprinkling and cooling by air, water sprinkling offered a lower temperature.

The open circuit voltage (V_{oc}), short circuit current (I_{sh}), and output power of different cooling systems were compared. Figure

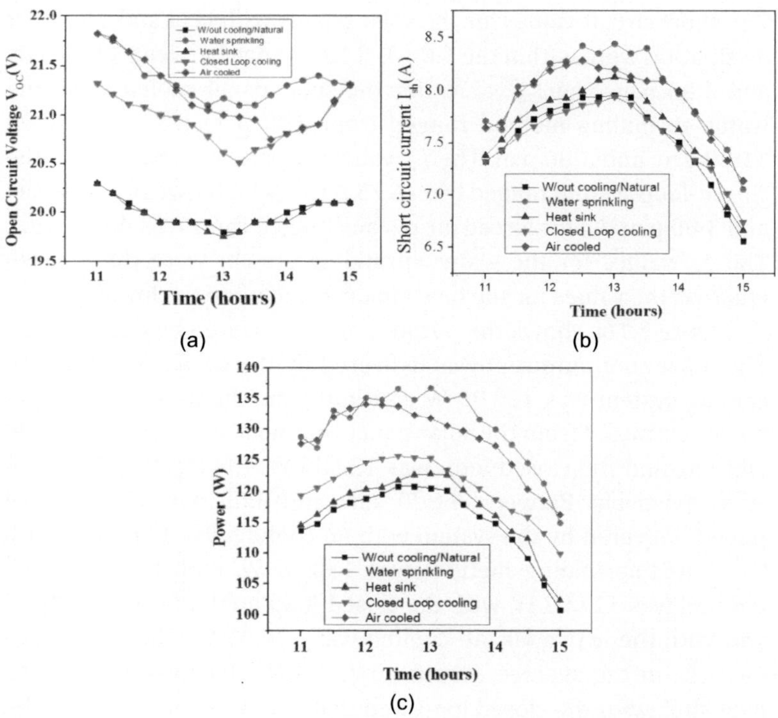

Figure 2.17 Comparative analysis of proposed cooling systems: (a) open circuit voltage (V_{oc}), (b) short circuit current (I_{sh}), and (c) output power.

2.17a shows that the open circuit voltage decreases with increasing temperature. The values for V_{oc} are higher at 11:00 a.m. and decreased with increasing temperature; the lowest values for V_{oc} were observed between 1:00 p.m. and 2:00 p.m., which then increased with decreasing temperature. Moreover, the values of V_{oc} for water sprinkling and air-cooled methods were higher (20.9 V to 21.82 V) than the V_{oc} for the system with no cooling and system with a heat sink (19.75 V to 21.18 V). V_{oc} for the closed-loop method was in the 20.51 V–21.32 V range.

The short circuit current (I_{sh}) curves for all the systems are shown in Fig. 2.17b. The short circuit values for the solar panel with no cooling ranged from 6.56 A to 7.96 A between 11:00 a.m. and 3:00 p.m., with maximum values between 12:30 and 1:30 p.m.

The short circuit values for the solar panels with heat sinks for heat dissipation were within the 6.66 A–8.13 A range between 11:00 a.m. and 3:00 p.m. The I_{sh} values for the solar panel cooled using the water sprinkling method ranged from 7.05 A to 8.44 A between 11:00 a.m. and 3:00 p.m. The I_{sh} values for the solar panel with the closed-loop system ranged from 6.73 A to 7.94 A between 11:00 a.m. and 3:00 p.m. I_{sh} for forced air by fans ranged from 7.13 A to 8.21 A. The I_{sh} values for the water sprinkling system were the highest, whereas the values for the heat sink method were the lowest.

Figure 2.16c shows the output power delivered by each system. The maximum output power delivered by the solar panel with no cooling system was 120.90 W at 12:30 p.m. The maximum output power extracted from the solar panel with heat sinks, the air-cooled method, and the closed loop was 122.86 W, 134.08 W, and 125.66 W, respectively. Between 11:00 a.m. and 3:00 p.m., the average power delivered by the system with no cooling was 115.30 W, with the water sprinkling method was 130.56 W, with the heat sink method was 117.03 W, with the closed-loop method was 128.41 W, and with the active air fan cooling was 120.75 W. The percentage increase in the average output power of the PV module with the heat sink, with the closed loop, and that was air cooled was 1.41%, 5.36%, and 11.2%, respectively. Interestingly, the SPV module with the water sprinkling method for cooling showed an enhanced but variable output power (max. 136.77 W) due to the cooling effect of water. The percentage increase in the output power of the SPV module with water sprinkling for cooling was 13.14%. Moreover, this method not only helped improve the output power by lowering the SPV temperature but also helped overcome the lowering of power caused by the accumulation of dust on the solar panels.

2.6.3 Combined Effect of a Cooling System and a Reflector

Aluminum foil was most effective and least expensive. Therefore, aluminum foil of the same size as the solar panel was used as a reflector for all four cooling systems and an increase in power was noticed from 8:00 a.m. to 5:30 p.m. Figure 2.18 shows the output power of these solar panels with different cooling systems.

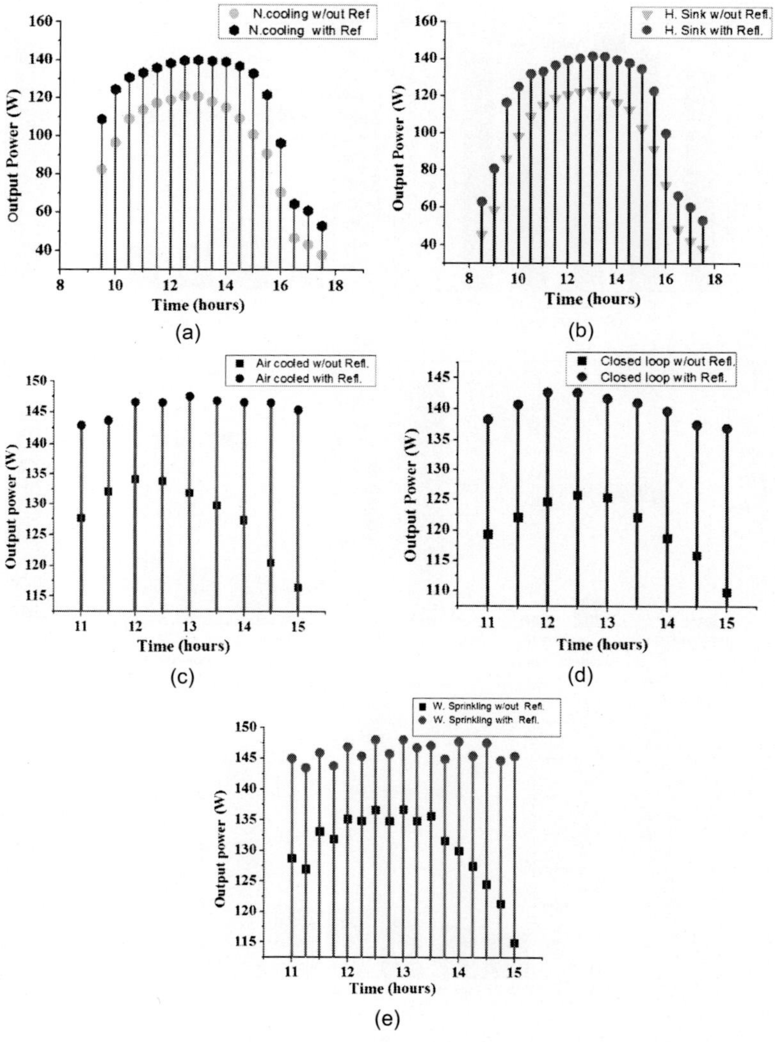

Figure 2.18 Combined effect of different cooling systems with reflectors on output power. (a) No cooling/natural cooling, (b) passive heat sink method, (c) active air fan cooling system, (d) closed-loop system, and (e) cooling by water sprinkling.

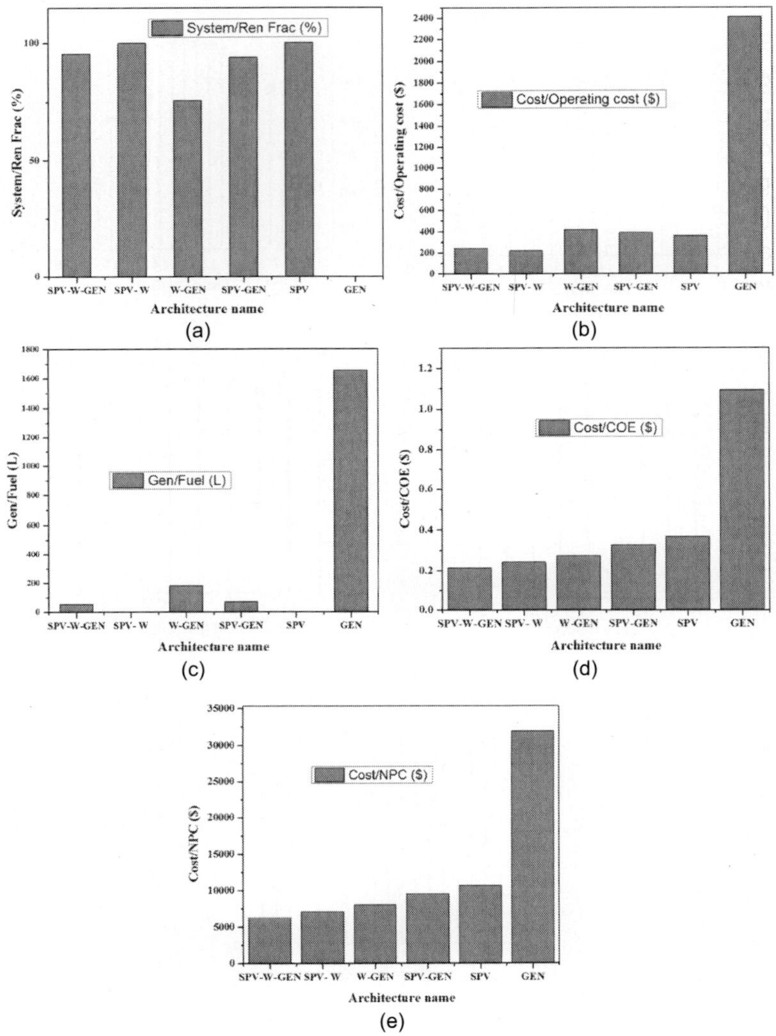

Figure 2.19 (a) Percentage renewable penetration, (b) operating cost ($), (c) fuel consumption, (d) cost of energy (COE), and (e) net present cost (NPC) for SPV-W-GEN, SPV, SPV-W, SPV-GEN, W-GEN, and GEN.

The use of aluminum foil with the cooling system helps obtain more power from the SPV system. The average power delivered by the solar panel with no reflector and no cooling system was 90.76 W. The average power delivered by the solar panel with no cooling system but with a reflector was 110.98 W. The average power delivered by the solar panel with a reflector and heat sink cooling was 113.25 W. The average power generated by the solar panel with a reflector and air cooling and cooling with a closed loop (with a reflector) was 122 W and 120 W, respectively. In contrast, the average power delivered by the solar panel with a reflector and the water sprinkling method for cooling was 125.75 W. The combined effect of using an Al foil reflector and the water sprinkling method for cooling offers a 38.55% increase in output power. The aluminum foils were more effective in the morning and evening and offered an average of approximately 48% increase in the output power of the SPV module.

2.6.4 Investigating the Appropriate Architecture/Structure

An assessment done by HOMER showed that SPV-W-GEN, W-GEN, and SPV-GEN offer 95.40%, 75.78%, and 93.86% renewable penetration, respectively, as shown in Fig. 2.19a. The operating costs of SPV-W-GEN, W-GEN, SPV-GEN, and SPV are \$239.86, \$413.81, \$388.90, and \$359.30, respectively (see Fig. 2.19b). The operating cost of SPV-W (\$218.952) was the least of all the systems considered. The fuel cost of all these systems was analyzed: SPV-W-GEN, W-GEN, SPV-GEN, and GEN used 52.42 L, 183.32 L, 69.92 L, and 1654.39 L of fuel, respectively, as illustrated in Fig. 2.19c. The COE for GEN is 1.09238, which is much higher than that for the rest. The COEs of SPV-W-GEN, SPV-W, W-GEN, SPV-GEN, and SPV are \$0.21467, \$0.24394, \$0.27541, and \$0.32599, respectively, as elaborated in Fig. 2.19d. The net present cost (NPC) for SPV-W-GEN is \$6259.86, for SPV-W is \$7110.53, for W-GEN is \$8031.25, for SPV-GEN is \$9505.97, for SPV is \$10638.53, and for GEN is \$31854.57, as shown in Fig. 2.19e. SPV-W, which has low operating cost and no fuel consumption, 100% renewable penetration, and low COE, was selected as a good and feasible HES, considering the weather conditions of the place where the experiments were performed (Arusha, Tanzania).

2.6.5 Cost-Effective Solution, Optimal Control, and Implementation Plan

A comparative analysis of different cooling systems with and without reflectors revealed the water sprinkling method to be the most efficient of all the methods assessed. On the other hand, every method has some advantages and disadvantages. For example, installing a heat sink on the rear side of a solar panel offers the least improvement in the output power of the SPV module but it does not require external power to operate. It offers some cooling even when other cooling systems are not recommended to be operated for that time period, that is, from 9:30 to 11:00 and from 15:30 to 17:00 (Fig. 2.20a).

The net increase in the output power by the heat sink method is very less compared to the initial investment required to install these heat sinks at the rear side of each solar panel and may not be an economical solution for many users. The performance of the heat sink method can be improved by increasing the natural airflow. On the other hand, the availability of natural air with a high flow rate cannot be guaranteed during real-time practical installation because it depends on the climate of the location where the PV system is to be installed.

The closed-loop method not only offers a moderate enhanced output power from the solar panel but also increases the water temperature (an average of 2.66°C in the experimental setup); see Fig. 2.20b.

In addition, it requires continuous operation of the water pump motor to circulate water in the pipe attached to the solar panel on its rear side. Electric power is needed to run this motor (a 10 W water pump was used for the closed-loop system in this experiment). However, very good SPV cooling and heated water generation are possible if both systems are combined together as shown in Fig. 2.21.

The air-cooled method is also very effective in reducing the temperature of the solar panel and enhancing the output power.

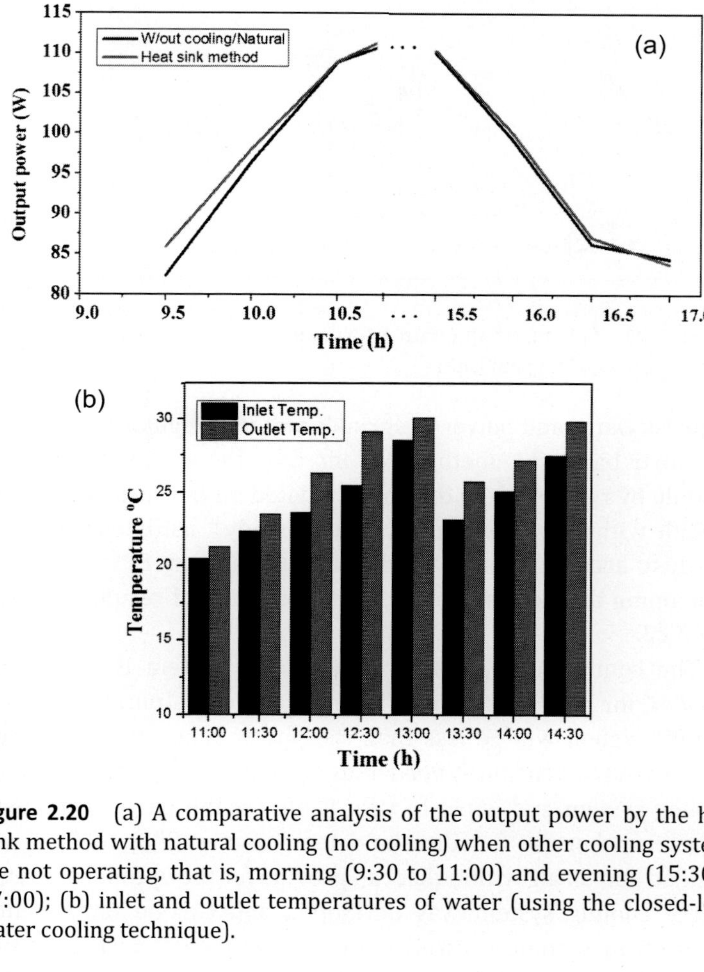

Figure 2.20 (a) A comparative analysis of the output power by the heat sink method with natural cooling (no cooling) when other cooling systems are not operating, that is, morning (9:30 to 11:00) and evening (15:30 to 17:00); (b) inlet and outlet temperatures of water (using the closed-loop water cooling technique).

This method, however, requires power to operate the fans attached to the solar panel for cooling. The performance of this method increases as the number of fans is increased. A total of eight fans, with each fan consuming 1.92 W (a total of 15.36 W), was required to operate this system. From the results, it is clear that a cooling system has a positive impact on the output power of the SPV system. On the other hand, selection of a specific cooling technique depends on the users' demand. The water sprinkling method is the most efficient of all the methods discussed in this chapter, but it

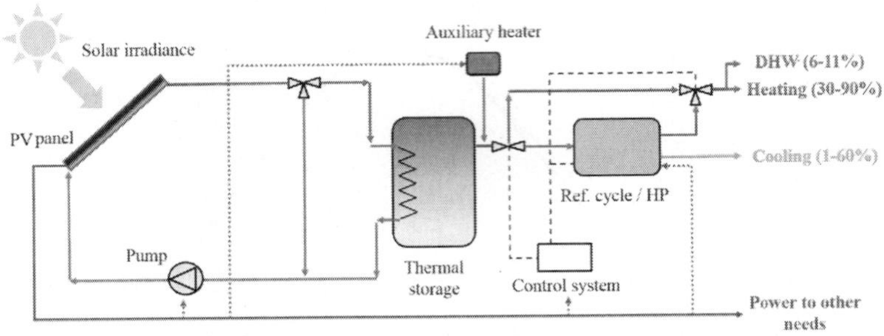

Figure 2.21 Schematic diagram of the proposed PV-T system for solar heating and cooling provision.[39]

requires water and power to sprinkle water at regular intervals. On the other hand, this method can increase the efficiency of the PV module by removing the dust accumulated on the surface of the PV module with passage of time. A detailed study[40] has been performed on dust accumulation and cleaning impact on SPV output by developing different scenarios, with summarized results shown in Fig. 2.22.

The continuous operation of a cooling system is not recommended for an SPV system. The increase in the output power of an SPV system will be less than the power required to run these systems. A mechanism is needed to determine when to turn on the cooling system and when to turn it off. On the basis of the work done elsewhere,[41,42] a threshold temperature (T_{th}) of 50°C was decided and an algorithm that can provide for the optimal operation of the cooling system was developed and can be implemented easily on an Arduino-controlled program with some basic electronic components. This algorithm describes that the cooling system should be turned on when $T_{\text{th}} = 50°\text{C}$ and turned off when the SPV module (T_{PV}) $= T_{\text{L}}$ or when $P_{\text{n}} > P_{n-1}$, that is, there is no improvement in output power using the cooling system. An algorithm for the optimal control of the BRPVS is shown in the flowchart in Fig. 2.23. Enhanced SPV output power requires an optimal BRPVS. The results in Section 2.4 show that the movement of reflectors varies with time and an optimal angle is required between the reflector and the solar panel. Figure 2.21 shows an algorithm for the timed-based optimal control of the BRPVS. If time

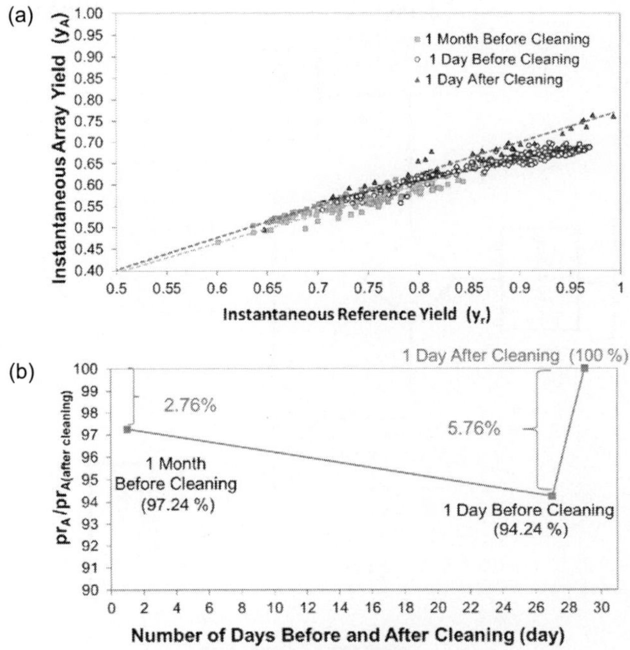

Figure 2.22 (a) Relationship between instantaneous array yield and instantaneous reference yield. (b) Relationship between instantaneous array performance ratio reduction (%) and the number of days before and after cleaning.[40]

$t = t_n$, turn both reflectors on; if $t < t_n$, move the right reflector described in the algorithm; if $t > t_n$, move the left reflector to an angle given in the algorithm. The algorithm shown in this figure summarizes only the movement of the reflector on the basis of the experimental studies in this experiment; an algorithm for hardware implementation may require a proper explanation of the system using a light-dependent resistor/cadmium sulfide sensor.[36]

The overall results of the research carried out in this chapter lead to a very novel HES consisting of an SPV system with a bireflector cooling system and a cooling system and wind power developed in the analysis of available resources in the region. An overview of hybrid PV-wind-ESS (composed of batteries) is shown in Fig. 2.24. Bireflector water heating system and cooling system are controlled using an energy management system.

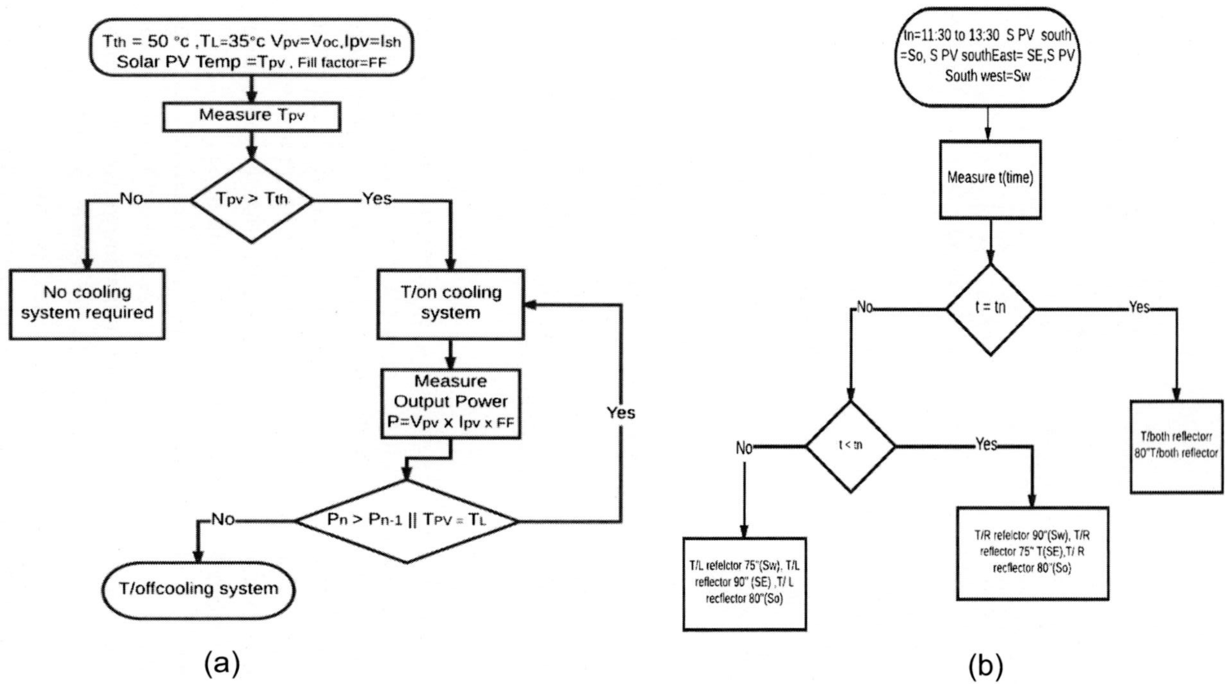

Figure 2.23 (a) Optimal control algorithm for a cooling system; (b) optimal control algorithm for the bireflector PV system (BRPVS).

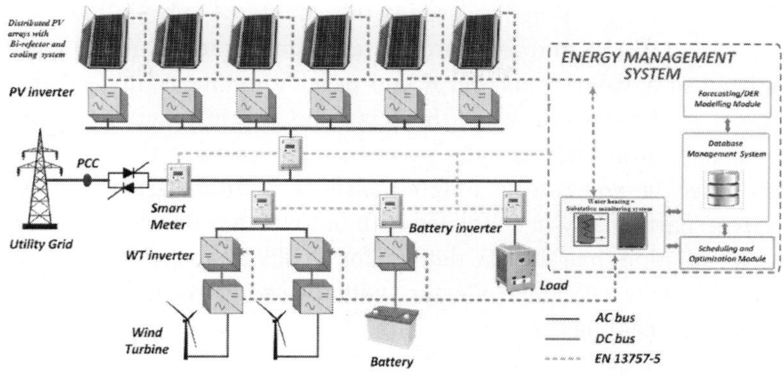

Figure 2.24 Novel hybrid photovoltaic/wind turbine/ESS microgrid with a BRPVS and a cooling controlling system.

2.7 Conclusions

The research presented in this chapter confirms the effectiveness of cooling and bireflector systems to improve the efficiency of existing PV systems. The HOMER software enables the design of a low-cost, high-efficiency HES using analysis performed at specific locations. Increase in the output power of the solar module depends upon the material used for the fabrication of the BRPVS. Silver-coated plane and spherical convex were less efficient and more expensive than the Al foil used as a reflector. Experimental results showed that the inexpensive Al foil as a reflector with the proposed BRPVS increases the output power from 20% to 35% or even more, which depends on the time of day and the position of the solar panel. This study assessed four different cooling systems, showing that the water sprinkling system was the most efficient (14%). The active-air-cooled fan system and the closed-loop system are comparatively more efficient than heat sink–based passive cooling, but both methods require continuous electric power for operation. The closed-loop system can be used as a source of water heating provided with some auxiliary heating system. The proposed BRPVS and water sprinkling cooling system can enhance the output power of the PV system by 22.8% to 38.6%. Continuous operation of any cooling system is not recommended, and the cooling system needs to follow the algorithm discussed in this chapter. A hybrid PV-wind-ESS with a BRPVS and an appropriate cooling system can easily be utilized to develop a microgrid and provide uninterrupted power supply to the location were these systems are installed. A highly enhanced output power solar system as discussed in this study is cost effective and can be adopted easily as a future green energy system, particularly in countries with hot weather or lower annual solar radiation than the STC defined for an SPV system. Future work should assess efficient reflector materials and low-input-power-consuming cooling systems.

References

1. J. M. Pearce (2002). Photovoltaics: a path to sustainable futures, *Futures*, **34**, pp. 663–674; K. Branker, M. J. M. Pathak, J. M. Pearce (2011). A review of solar photovoltaic levelized cost of electricity, *Renewable Sustainable Energy Rev.*, **15**, pp. 4470–4482.
2. C. Deline, A. Dobos, S. Janzou, J. Meydbray, M. Donovan (2013). A simplified model of uniform shading in large photovoltaic arrays, *Sol. Energy*, **96**, pp. 274–282.
3. D. Grimmer, K. Zinn, K. Herr, B. Wood (1978). Augmented solar energy collection using different types of planar reflective surfaces: theoretical calculations and experimental results, *Sol. Energy*, **21**, pp. 497–501.
4. D. McDaniels, D. Lowndes, H. Mathew, J. Reynolds, R. Gray (1975). Enhanced solar energy collection using reflector-solar thermal collector combinations, *Sol. Energy*, **17**, pp. 277–283.
5. S. C. Seitel (1975). Collector performance enhancement with flat reflectors, *Sol. Energy*, **17**, pp. 291–295.
6. L. T. Kosti, Z. T. Pavlovi (2012). Optimal position of flat plate reflectors of solar thermal collector, *Energy Build.*, **45**, pp. 161–168.
7. M. D. J. Pucar, A. R. Despic (2002). The enhancement of energy gain of solar collectors and photovoltaic panels by the reflection of solar beams, *Energy*, **27**, pp. 205–223.
8. M. Hall, A. Roos, B. Karlsson (2005). Reflector materials for two-dimensional low-concentrating photovoltaic systems: the effect of specular versus diffuse reflectance on the module efficiency, *Prog. Photovoltaics Res. Appl.*, **13**, pp. 217–233.

9. S. L. Grassie, N. R. Sheridan (1977). The use of planar reflectors for increasing the energy yield of flat-plate collectors, *Sol. Energy*, **19**, pp. 663–668.
10. B. Perers, B. Karlsson (1993). External reflectors for large solar collector arrays simulation model and experimental results, *Sol. Energy*, **51**, pp. 327–337.
11. T. Matsushima, T. Setaka, S. Muroyama (2003). Concentrating solar module with horizontal reflectors, *Sol. Energy Mater. Sol. Cells*, **75**, pp. 603–612.
12. M. A. Khan, S. Badshah (2014). Design and analysis of cross flow turbine for micro hydro power application using sewerage water, *RJASET*, **8**, pp. 821–828.
13. E. Chaniotakis (2001). Modelling and analysis of water cooled photovoltaics, M.Sc. thesis, Faculty of Energy System and Environment, Department of Mechanical Engineering, University of Strathclyde, Glasgow, Scotland.
14. S. Dubey, G. N. Tiwari (2008). Thermal modeling of a combined system of photovoltaic thermal (PV/T) solar water heater, *Sol. Energy*, **82**, pp. 602–612.
15. H. Batoul (2008). Flow simulation improves photovoltaic solar panel performance, *Technical Report*, Schueco International, Paris, France. <http://www.schueco.com/>.
16. J.-C. Wang et al. (2011). High-accuracy maximum power point estimation for photovoltaic arrays, *Sol. Energy Mater. Sol. Cells*, **95**(3), pp. 843–851.
17. J. K. Tonui, Y. Tripanagnostopoulos (2007). Improved PV/T solar collectors with heat extraction by forced or natural air circulation, *Renewable Energy*, **32**, pp. 623–637.
18. A. Kluth (2008). Using water as a coolant to increase solar panel efficiency, California State Science Fair, California, USA.
19. X. Tang, Z. Quan, Y. Zhao (2010). Experimental investigation of solar panel cooling by a novel micro-heat pipe array, *Energy Power Eng.*, **2**, pp. 171–174.
20. M. C. Alonso Garcia, J. L. Balenzategui (2004). Estimation of photovoltaic module yearly temperature and performance based on nominal operation cell temperature calculation, *Renewable Energy*, **29**, pp. 1997–2010.
21. Y. A. Cengel, M. A. Boles (2005). *Thermodynamics: An Engineering Approach*, 5th ed., McGraw-Hill Science, New York.

22. M. Ibrahim, B. Zinber, H. El-Sherif, E. Hamouda (2009). Advanced photovoltaic test park in Egypt for investigating the performance of different module and cell technologies, in *Proceedings of 24th Symposium Photovoltaic Solar Energy*, Bad Staffelstein, Germany.
23. B. Zinber, G. Makrides, W. Schmitt, G. E. Georghiou, J. H. Werner (2007). Annual energy yield of 13 photovoltaic technologies in Germany and in Cyprus, in G. Willeke, H. Ossenbrink, P. Helm (eds.), *Proceedings of 22nd European Photovoltic Solar Energy Conference*, Munich, Germany.
24. A. Moehlecke, F. S. Febras, I. Zanesco (2013). Electrical performance analysis of PV modules with bifacial silicon solar cells and white diffuse reflector, *Sol. Energy*, **96**, pp. 253–262.
25. M. Pavlov, et al. (2015). Experimental and numerical study of the influence of string mismatch on the yield of PV modules augmented by static planar reflectors, *IEEE J. Photovoltaics*, **5**(6), pp. 61686–1691.
26. J. A. Duffie, W. A. Beckman (2013). *Solar Engineering of Thermal Processes*; John Wiley & Sons, Hoboken, NJ, USA.
27. L. T. Kostic, T. M. Pavlovic, Z. T. Pavlovic (2010). Optimal design of orientation of PV/T collector with reflectors, *Appl. Energy*, **87**, pp. 3023–3029.
28. U. Stutenbaeumer, B. Mesfin (1999). Equivalent model of monocrystalline, polycrystalline and amorphous silicon solar cells, *Renewable Energy*, **18**, pp. 501–512.
29. H. M. Bahaidarah, B. Tanweer, P. Gandhidasan, S. Rehman (2015). A combined optical, thermal and electrical performance study of a V-trough PV system—experimental and analytical investigations, *Energies*, **8**(4), pp. 2803–2827.
30. C. Schlick (1994). An inexpensive BRDF model for physically-based rendering, *Comput. Graphics Forum*, **13**, pp. 233–246.
31. R. W. Andrews, P. Andrew, M. P. Joshua (2015). Photovoltaic system performance enhancement with nontracking planar concentrators: experimental results and bidirectional reflectance function (BDRF)-based modeling, *IEEE J. Photovoltaics*, **5**, pp. 1626–1635.
32. M. Hall, A. Roos, B. Karlsson (2005). Reflector materials for two-dimensional low-concentrating photovoltaic systems: the effect of specular versus diffuse reflectance efficiency on the module, *Prog. Photovoltaics Res. Appl.*, **13**, pp. 217–233.
33. M. A. Khan, B. Ko, E. Alois Nyari, S. E. Park, H.-J. Kim (2017). Performance evaluation of photovoltaic solar system with different cooling methods

and a bi-reflector PV system (BRPVS): an experimental study and comparative analysis, *Energies*, **10**, p. 826.

34. U. J. Rajput (2016). Application of natural convection for photovoltaic cooling and photocatalytic disinfection, *Electronic Thesis and Dissertation Repository*, 3671, https://ir.lib.uwo.ca/etd/3671.
35. C. G. Popovici, et al. (2016). Efficiency improvement of photovoltaic panels by using air cooled heat sinks, *Energy Procedia*, **85**, pp. 425–432.
36. N. Hartmann, C. Glueck, F. Schmidt (2011). Solar cooling for small office buildings: comparison of solar thermal and photovoltaic options for two different European climates, *Renewable Energy*, **36**(5), pp. 1329–1338.
37. A. R. Amelia, et al. (2016). Cooling on photovoltaic panel using forced air convection induced by DC fan, *Int. J. Electr. Comput. Eng.*, **6**, pp. 526–534.
38. S. Odeh, M. Behnia (2009). Improving photovoltaic module efficiency using water cooling, *Heat Transfer Eng.*, **30**, pp. 499–505.
39. IEA (2012). Technology roadmap solar heating and cooling. Tech Rep. Paris: International Energy Agency, <http://www.iea.org/publications/freepublications/publication/Solar_Heating_Cooling_Roadmap_ 2012_WEB.pdf>
40. N. Sakarapunthip, et al. (2017). Effects of dust accumulation and module cleaning on performance ratio of solar rooftop system and solar power plants, *Jpn. J. Appl. Phys.*, **56**.
41. M. A. Khan, S. Badshah (2014). Design and analysis of cross flow turbine for micro hydro power application using sewerage water, *RJASET*, **8**, pp. 821–828.
42. J. Barrau, A. Perona, A. Dollet, J. Rosell (2014). Outdoor test of a hybrid jet impingement/micro-channel cooling device for densely packed concentrated photovoltaic cells, *Sol. Energy*, **107**, pp. 113–121.
43. K. A. Moharram, M. S. Abd-Elhady, H. A. Kandil, H. El-Sherif (2013). Enhancing the performance of photovoltaic panels by water cooling, *Ain Shams Eng. J.*, **4**, pp. 869–877.

Chapter 3

A Blended SPS-ESPS Control DAB-IBDC for a Stand-Alone Solar Power System

In sustainable energy applications, stand-alone solar power systems are mostly preferred for self-powered energy zones. In all stand-alone renewable power systems, batteries are still preferred as the common energy storage device. On the other hand, batteries are not applicable for high-peak-power-demand applications because of their low power density. A supercapacitor is a preferable high-power-density energy storage device for high-peak-power applications. A 2 kW, 50 kHz digital control dual-active bridge-isolated bidirectional DC–DC converter (DAB-IBDC) was developed for interfacing the supercapacitor bank in a stand-alone solar power system. This chapter proposes a blended SPS-ESPS digital control algorithm for a DAB-IBDC instead of using a traditional single-phase-shift (SPS)-control algorithm, which is commonly used for applications with large input to output voltage variations. This proposed blended SPS-ESPS control algorithm achieved high power conversion efficiency during a large input to output voltage variation, over a traditional phase-shift-control algorithm by reducing the backflow power and the current stress in a circuit. This system also achieved maximum power point for solar modules and enhanced

Solar Power and Energy Storage Systems
Hee-Je Kim

ISBN 978-981-4800-13-6 (Hardcover), 978-0-429-45877-4 (eBook)
www.panstanford.com

rapid charging-discharging for a supercapacitor bank. Both SPS and blended SPS-ESPS control algorithms were verified experimentally using a 2 kW DAB-IBDC topology implemented with a stand-alone power system that combined a 2000 W input solar module and a 158 Wh supercapacitor bank.

3.1 Introduction

Over the last decade, fossil fuel technology has become less attractive due to the depletion of reserves on earth. Renewable energy sources are believed to be an alternative energy source to replace fossil fuels. Among renewable energy systems, stand-alone photovoltaic (PV) systems are commonly preferred in remote areas. A typical stand-alone system incorporates a PV panel, a maximum power point tracker (MPPT), an energy storage system, and a charging controller.[1] A PV system is not an ideal DC source for charging energy storage systems gradually because it does not satisfy the desirable requirements of storage devices. The output of such a system is dependent on the frequently changeable weather conditions, which are largely unpredictable. Therefore, the PV system output is unreliable and an optimal charge-discharge cycle cannot be guaranteed.[2] With lead acid batteries, which are the most preferred common energy storage devices in stand-alone renewable energy systems, this unreliable charging results in a battery in a low charge state, which leads to sulfating and stratification, both of which shorten battery life cycle.[3] To avoid this, an MPPT and a charge controller are used as an intermediate circuit in a renewable solar energy system between the solar source and the energy storage device.

Recently, researchers have focused on supercapacitors as energy storage devices instead of using batteries. Supercapacitors have several advantages over batteries, such as the ability to handle high peak power, high power density, instantaneous turn-on ability, and rapid charge-discharge capability. These unique properties of a supercapacitor make it an ideal energy storage device in certain load applications where there is a high peak power demand. For example, in motor starting applications, the starting current requirement can be 6–10 times higher than the normal operating current of the motor. Electric vehicles are likely to employ electrically powered actuators for high-power transient loads. A supercapacitor-based energy storage system has been proposed to meet the peak power demands of the electric vehicle loads and absorb the regenerated power.[4,5] On the other hand, the major limitation of supercapacitors is their lower energy density, so they cannot store large amounts of energy the way a battery does. For example, if a supercapacitor is to be used to store the same amount of energy as a lead acid battery, with the present technology, the supercapacitor must be 10 times larger than the corresponding battery. Although a large size is considered a major disadvantage in a supercapacitor, it can still be applied in an energy storage system because of its higher power density. Many studies of the material side have attempted to improve the supercapacitor energy density to make it an efficient energy storage device.[6] In the future, the energy density of supercapacitors will hopefully be much improved compared to that of conventional batteries. Nowadays, lead acid batteries with a solar source are commonly used for developing self-powered energy zones commonly known as no-power grid zones. Although lead acid batteries are appropriate for self-powered energy zones, while handling high-peak-power appliances, such as air conditioners, mixers, and grinders, batteries face enormously increased current stress, which leads to a lot of damage to the batteries. Many researchers have proposed supercapacitors or combinations of supercapacitors and batteries as energy storage devices in stand-alone power systems for developing a better self-powered energy zone.[5]

Many DC–DC converter topology and control approaches are being proposed in stand-alone renewable energy storage systems for achieving maximum power and high power conversion efficiency and secure fast charging-discharging.[7–10] In the design of high-power DC–DC converters, the main trade-offs are between power efficiency and power converter. In the 1970s, the power semiconductor technology was developed, resulting in a significant increase in the power density of power electronic converters. With more progressive power semiconductor devices, switch mode power electronic converters with a higher switching frequency have been developed. Therefore, the volumes of the converter circuits can be decreased. Power semiconductors and integrated circuits have been developed to economically produce DC–DC converters.

On the basis of reviews, from among all the DC–DC converters for interfacing renewable sources and high-power-density energy storage devices, because of its unique characteristics a dual-active bridge-isolated bidirectional DC–DC converter (DAB-IBDC) was proposed by the authors in the early 1990s.[11] On the other hand, because of the performance limitations of power devices, the power losses of a DAB-IBDC were high and the efficiency was unacceptable. In recent years, the advances in new power devices and magnetic materials have made a DAB-IBDC a simple and low-loss converter circuit compared to all other converters.[12] For power conversion, the DAB-IBDC topology is popular among researchers because of its high performance, high efficiency, galvanic isolation, ability for handling high power density, both-side power flow ability, and inherent soft switching property. These features make the DAB-IBDC a typical circuit for high-power-density applications.[13] Several papers have been published, and the performance and comprehensive analyses of DAB converters have been reported.

Design considerations of the DAB converter and high-frequency DAB transformer were discussed briefly in terms of future trends.[14] Naayagi et al. proposed a new model for the steady-state analysis of a DAB converter circuit that produces equations for rms and average device currents and rms and peak inductor currents. These equations are useful for predicting the losses that occur in the device and passive components.[15] A previous study[16] optimized the design of a dual-active bridge converter, permitting flexible interfacing to energy storage devices to achieve uninterrupted power supply in military applications. The DAB performance for the next-generation power conversion systems (PCSs) using ultracapacitor-based technologies has been validated.[17] The performance of an ultracapacitor-based DAB converter dynamic was modeled and analyzed.[18] According to previous research based on a digital control algorithm, extended-single-phase-shift (ESPS) control of isolated bidirectional DC–DC converters is improving system efficiency of the circuit for power distribution between energy source and energy storage system in microgrids over a wide input voltage and output

Figure 3.1 Block diagram of a stand-alone photovoltaic power system.

power range.[19] The current stress and backflow power of an isolated converter circuit was analyzed using ESPS control, and it was verified that ESPS achieves greater efficiency with an inner phase limitation over single-phase-shift (SPS) control under microgrid applications.[20]

This chapter proposes a modified digital control DAB-IBDC to be deployed in a stand-alone solar power system coupled with a high-power-density energy storage device, as shown in Fig. 3.1. The proposed algorithm can operate the solar source at its maximum power point (MPP), converting power with the maximum conversion efficiency and enhancing the rapid charging-discharging for high-power-density energy storage devices. In the digital control algorithm, the blended SPS-ESPS control algorithm operates a converter with the maximum power conversion efficiency by reducing the backflow power in a converter circuit. In such large-voltage-variation applications ($k > 1$), like a renewable energy source, the backflow power of a DAB-IBDC is large, which leads to an increase in the current stress in a circuit. Therefore, the overall circuit performance will be affected and the power conversion efficiency will be reduced. The proposed blended SPS-ESPS digital control algorithm continuously monitors the backflow power in a circuit and reduces it as much as possible by adjusting the inner phase shift of the primary bridge. In this laboratory, a 2 kW DAB-IBDC topology with metal-oxide-semiconductor-field-effect-transistor (MOSFET) switching bridges was developed. The overall control algorithm was verified experimentally using this topology in a 2000 W stand-alone solar power system. To analyze the modified control algorithm, the experiment was also conducted using the traditional SPS algorithm. The results were discussed based on the experimental results, and it was concluded that the proposed blended SPS-ESPS digital control DAB-IBDC gains more efficiency over traditional phase-shift DAB converters. The results showed that the blended SPS-ESPS digital control DAB-IBDC is more suitable for a stand-alone renewable power system in high-peak-power-demand applications, such as electric vehicles and motors.

3.2 Principle of the DAB-IBDC Circuit

Figure 3.2 presents a schematic diagram of a DAB-IBDC circuit. These circuits are simply structured, with two full bridges isolated by a high-frequency high-power transformer. Both-side power flow exists in a circuit and is controlled by the DSP controller (TMS320F28335, Texas Instruments, Dallas, TX, USA). The schematic diagram, principle of operation, and steady-state model of a circuit are explained below.

A dual-active bridge converter contains an isolated transformer with a leakage inductance, L_{leq}. The required inductance for the circuit is provided partly or entirely by the transformer leakage inductance. If needed, an additional coupling inductor may be connected in series with the transformer leakage inductor to boost the overall inductance of the converter circuit depending on the conversion requirements. The isolated transformer isolates the two full-bridge circuits, in that the primary-side full bridge is connected to a high-voltage DC source, called a high-voltage bridge (V_{ab}), and another full bridge, on the secondary side, is connected to a low-voltage energy storage system called a low-voltage bridge (V_{cd}). In forward mode, the current flows from the source to the supercapacitor until the supercapacitor is charged fully. The load is connected to the primary side of the circuit because it can be powered up by the energy storage system in the backward mode of operation. The full bridges on both sides consist of four 47N60C3 MOSFETs (Infineon Technologies, Hong Kong, China), as shown in Fig. 3.2, and are controlled by a high-frequency square wave voltage given by the DSP followed by a MOSFET driver. For bridges on both

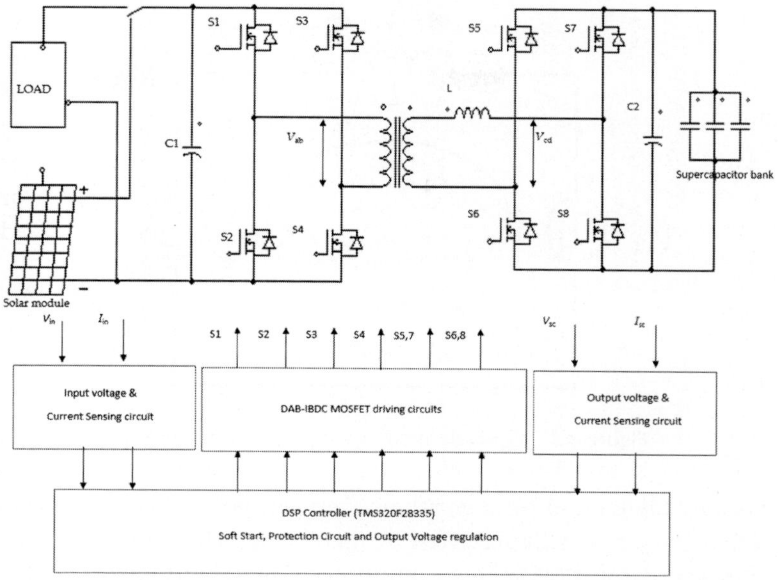

Figure 3.2 Schematic diagram of a digital control dual-active bridge-isolated bidirectional DC–DC converter (DAB-IBDC).

sides, two square waves can be suitably phase-shifted with respect to each other to control the power flow direction. In forward mode, the phase of the V_{cd} square wave is shifted by the phase of the V_{ab} square wave. In contrast, in backward mode, the phase of the V_{ab} square wave is shifted by the phase of the V_{cd} square wave. Therefore, bidirectional power flow is enabled in a DAB converter. Current flow occurs because of the voltage difference across the inductor due to the phase shift between these two bridges' square wave voltages.

3.2.1 Equivalent Circuit of Phase-Shift Control

In Fig. 3.3, L is the sum of the transformer leakage inductance, L_{leq}, and coupling auxiliary inductance L_1. V_{ab} and V_{cd} are the equivalent AC output voltages of the primary and secondary sides, respectively. V_L and I_L are the voltage and current of the inductor, L. The power flow direction and magnitude of power can be controlled simply by adjusting the phase shift between V_{ab} and V_{cd}. Although

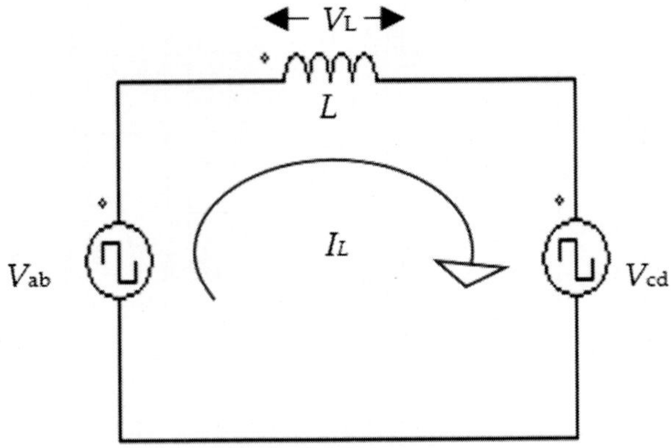

Figure 3.3 Equivalent circuit of phase-shift control.

various digital control algorithms were proposed, the traditional SPS algorithm is still preferred by many researchers for its unique characteristics, such as high power handling ability and handling large input to output voltage variations.[2] ESPS control is preferred by researchers to increase the conversion efficiency by reducing the backflow power and the current stress in a circuit. Regarding increasing conversion efficiency, compared to other digital control algorithms it can be implemented easily by adding an inner phase shift at the primary side bridge. Both the control algorithm and the forward mode waveform shown in Fig. 3.4 were assessed by waveform steady-state analysis.

3.2.2 Steady-State Analysis

Figure 3.4 shows the main waveforms of a DAB-IBDC in SPS and ESPS control, where T_{hs} is a half switching period and d is the phase-shift ratio between the primary and secondary voltages of the isolation high-frequency transformer, where $0 \leq d \leq 1$. In Fig. 3.4a V_{ab} and V_{cd} are both square wave voltages and the interaction between these two voltages occurred through the inductor of the circuit. Therefore, the primary voltage is always in the miss-phase with the primary current because it passes through the inductor. The average currents of the leakage inductor are derived based on the waveform I_L. The difference in voltage between V_{ab} and

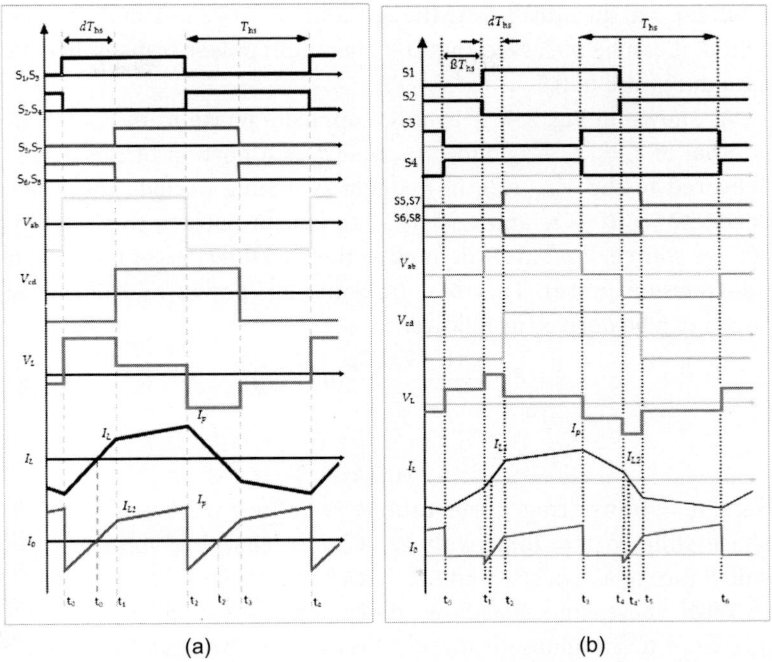

Figure 3.4 Key operating waveforms of a DAB-IBDC during forward (buck) mode. (a) Traditional single-phase-shift (SPS) control, and (b) extended-single-phase-shift (ESPS) control.

V_{cd} appears across the inductor, L, and inductor current, I_L, at the switching instances, t_1 and t_2, respectively:

$$I_p = \frac{T_s}{4L}[nV_{in} + V_0\,(2d - 1)] \tag{3.1}$$

$$L_1 = \frac{T_s}{4L}[nV_{in}\,(2d - 1) + V_0] \tag{3.2}$$

The waveform is a periodic over half a cycle. Thus, dividing the area by the duration $T_s/2$, we get the average output current of the DAB converter:

$$I_0 = \frac{nV_{in}\,T_s}{2L}(d - d^2) \tag{3.3}$$

Normally, the average output current depends on $\frac{nV_{in}\,T_s}{2L}$:

$$I_0' = (d - d^2) \tag{3.4}$$

From Eq. 3.4, on substituting the full control range 0–1 in the phase shift d, it can be observed that the maximum power transfer occurs for a duty ratio of 0.5.

As shown in Fig. 3.4, I_L is of the opposite phase from V_{ab} for an interval of $t = t_0 - t_0'$ and $t = t_2 - t_2'$ is a portion of the power delivered to the V_{out} side in a single switching period. The other portion $t = t_0' - t_1$ and $t = t_2' - t_3$, is sent back to the primary voltage source, V_{in}. This is defined as the backflow power for a given transmission power. The total transmission power and backflow power can be derived as follows:

$$P = \frac{nV_{in}V_{out}T_s}{2L}(d - d^2) \tag{3.5}$$

$$P_{bf} = \frac{nV_{in}V_{out}T_s[k + (2d - 1)]^2}{16L(k + 1)} \tag{3.6}$$

Here, n = transformer turns ratio, k = voltage transfer ratio, d = phase shift, V_{in} = input voltage, V_{out} = charging voltage, L = inductance of a converter, and T_s = switching time.

With increasing backflow power, the forward power also increases to compensate for the loss caused by backflow power. The circulating power and current stress are then increased, which results in great loss in power devices and magnetic components and low efficiency of the converter:

$$I_{max} = \frac{nV_2T_s}{4L}(2d - 1 + k) \tag{3.7}$$

From Eqs. 3.1 to 3.7, maximum power transfer occurs at a duty ratio of 0.5. Even the same transfer power is achieved for a duty of 0.5–1, which is the same as 0 to 0.5, in that the backflow power and the current stress will be low in a region where the duty is 0–0.5 compared to the duty 0.5–1 region.

To decrease the backflow power, V_{ab} should not be confined to a square wave with a 50% duty cycle. In Fig. 3.4b, S_1 and S_4 have a phase shift of β, which reduces the V_{ab} square wave duty cycle. The transformer primary voltage will emerge as three levels instead of two levels in SPS control. This alters the behavior of the I_L current. The inductor current (I_L) changes in three instances of time, t_2, t_3, and t_4, which are denoted as I_{L1}, I_P, and I_{L2}, respectively:

$$I_{L2} = \frac{T_s}{4L}[nV_{in}(1 - \beta) + V_0\,(2d - 1)] \tag{3.8}$$

$$I_{L1} = \frac{T_s}{4L}[nV_{in}(2d - 1 + \beta) + V_0] \tag{3.9}$$

$$I_p = \frac{T_s}{4L}[nV_{in}(1 - \beta) + V_0(2d - 1 + 2\beta)] \tag{3.10}$$

Here, β = inner phase shift, d = outer phase shift, and T_s = switching time.

If you see a waveform, it's a periodic over half a cycle. Dividing the area by the duration $T_S/2$ gives the average output current of the DAB converter as:

$$I_0 = \frac{nV_{in}T_s}{2L}\left[d - d^2 - dß + \frac{\beta}{2} - \frac{\beta^2}{2}\right] \tag{3.11}$$

Normalizing the average output current on the basis of the values $\frac{nV_{in}T_s}{2L}$ gives:

$$I_0' = d - d^2 - d\beta + \frac{\beta}{2} - \frac{\beta}{2} \tag{3.12}$$

The backflow power appearance times are divided into two intervals $t = t_0 - t_1$, $t = t_1 - t_1'$ and $t = t_3 - t_4$, $t = t_4 - t_4'$. The backflow power is zero as the transformer primary voltage reaches zero for the first intervals. Thus, the overall backflow power decreases for a given transmission power:

$$P' = \frac{1}{T_{hs}}\int_0^{T_{hs}} V_{h1}i_L(t)\,dt = \frac{nT_sV_1V_2}{2L}(d(1-d) + \frac{1}{2}\beta(1 - \beta - 2d) \tag{3.13}$$

Backflow power:

$$P_{bf}' = \frac{1}{T_{hs}}\int_{t1}^{t'1} V_{h1}|i_L(t)|\mathrm{d}t = \frac{nT_sV_1V_2[k(1-\beta) + (2d-1)]^2}{16L(k+1)}, \tag{3.14}$$

where, k = voltage transfer ratio.

Current stress under ESPS control:

$$I'_{max} = \frac{nT_sV_2}{4L}(k(1-\beta) + (2\beta + 2d - 1), \tag{3.15}$$

where $i_L(t_1) < 0$; from the equation, we have

$$k > \frac{1 - 2d}{1 - \beta}. \tag{3.16}$$

When $k \leq (1 - 2d)/(1 - \beta)$, the backflow power is zero.

In Fig. 3.4b, β is the phase-shift ratio between the driving signals of S1 & S4 and S2 & S3 in the primary bridge. β is defined as the inner phase-shift ratio, where $0 \leq \beta \leq 1$. d is the phase-shift ratio between the primary and secondary voltages of the isolation transformer. d is often known as the outer phase-shift ratio, where $0 \leq d \leq 1$ and $0 \leq \beta + d \leq 1$. Compared to SPS control, ESPS control has not only an outer-phase-shift ratio but also an inner-phase-shift ratio, which will decrease the current stress, expand the regulating range of transmission power, and enhance the regulating flexibility.

3.3 Experimental Setup

Figure 3.5 presents the developed 2 kW DAB-IBDC circuit and experimental setup. In the DAB-IBDC topology, active bridges on both sides have been designed with a 47N60C3 power MOSFET. The equivalent gate capacitance is expressed by the gate-source and gate-drain capacitors. The drain source resistance is 0.07 Ω, which is considered the best value among all other MOSFET switches. According to the datasheet, the drain source voltage is nearly 650 V, the drain current is 47 A, and the pulsed drain current is 141 A. To trigger the MOSFET switches at a high frequency ($f > 10$ kHz),

Figure 3.5 Developed DAB-IBDC and experimental setup.

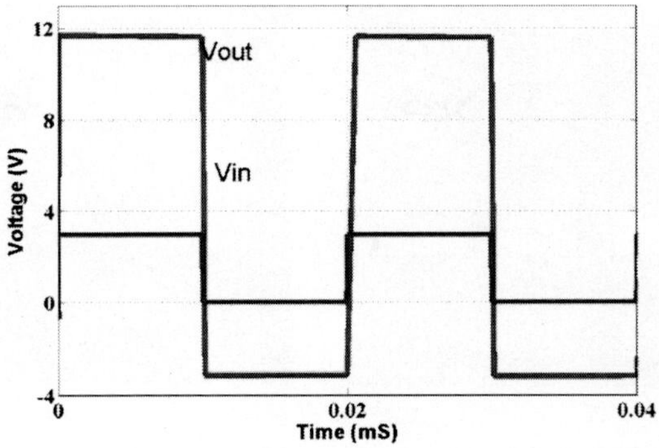

Figure 3.6 Input and output signal waveform of a MOSFET driver circuit.

pulse-width-modulation (PWM) signals (±7 to ±30 V, 3 mA) need to be applied across the gate-source terminal.

The applied PWM signals are generated by a digital signal processor (TMS320F28335) on the basis of the control algorithm. As these PWM signals from DSP do not have sufficient power to operate switches, a MOSFET driving circuit was designed and used to interface the DSP and MOSFET switches. The driver circuit input and output waveform signal are shown in Fig. 3.6. The designed MOSFET driver converts the low-power 0 V to 5 V PWM signal to a −3 V to 12 V high-power PWM signal, which can turn the switches on and off.

The previous PCS is employed mainly with a line frequency (LF) transformer to achieve isolation and voltage matching. Basically, the LF transformer size is extremely large, which hinders the design of a simple PCS. In the LF transformer, current-voltage distortions and losses are high due to core saturation. Overall, in previous PCSs, the LF transformer reduced the power conversion efficiency and power density. With the aim of increasing the efficiency and power density of a converter, a high-frequency PCS has been proposed recently. When the switching frequency is above 20 kHz, the PCS's noise can be reduced greatly and the size of the magnetic materials, such as a transformer or an inductor, is also reduced. This creates a path to designing a simple and lossless PCS. As a 50 kHz DAB-IBDC is preferred, a simple high-power high-frequency transformer was

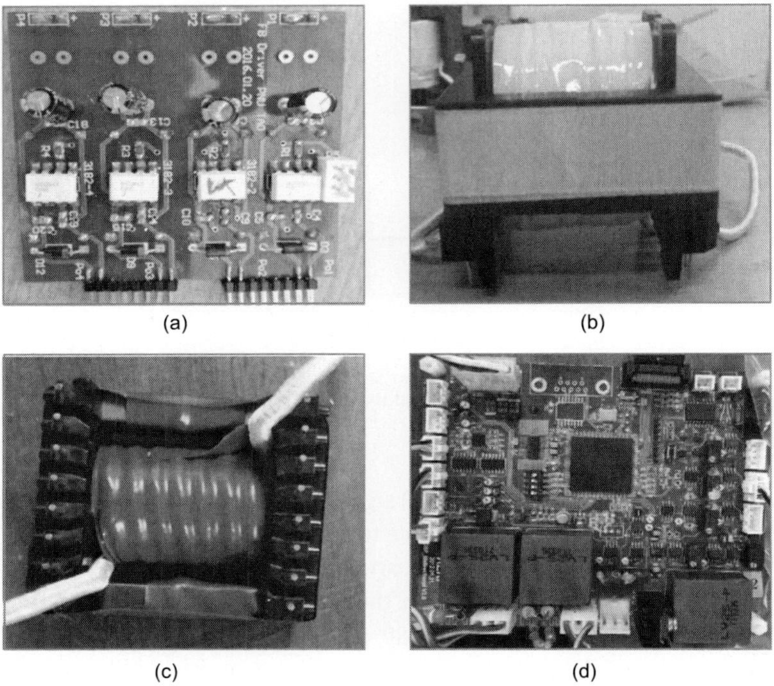

Figure 3.7 DAB-IBDC subcircuits and magnetic components. (a) MOSFET driving circuit; (b) transformer; (c) inductor; (d) digital controller.

designed using a ferrite core, as shown in Fig. 3.7, with a transform turn ratio (n) of 3:4 and a leakage inductance of 5.26 μH.

According to the DAB-IBDC design, the required inductance is 9.98 μH, which can be given by the transformer leakage inductance, either alone or combined with an external coupling inductor. As per the design, to reach the required inductance of the circuit, at the transformer secondary side, an additional inductor with a value of 4.72 μH is connected in series. The additional inductor is called the coupled inductor, and the overall inductance is called the auxiliary inductance of the circuit. Table 3.1 lists the specifications of the developed 2 kW DAB-IBDC. The input source of the DAB converter was a 220 V renewable solar energy DC source, which was obtained from six 38 V, 340 W solar panels connected in series. The output was connected to the supercapacitor bank, which were

Table 3.1 DAB-IBDC design specifications

Parameter	Rating
Input voltage	200–220 V
Power	2000 W
Output voltage	48 V
Transformer turns ratio	3:4
Switching frequency	50 kHz
Secondary leakage inductance	5.26 μH
Auxiliary inductance	4.72 μH
Input capacitor	1200 μF
Output capacitor	1800 μF

built with three 48 V, 165 F, and 53 Wh Maxwell supercapacitors connected in parallel. The overall supercapacitor bank specification was 48 V, 495 F, and 159 Wh. The proposed DAB-IBDC was controlled by the digital control board, which was designed using a DSP (TMS320F28335) IC along with many subcircuits, such as short circuit protection, PWM protection, soft start, voltage sensor, and current sensor, shown in Fig. 3.7d.

3.4 Digital Control System

A closed-loop controller is required in power converters for regulating the output voltage/current, and it is also needed to compensate the source/load disturbance. The closed-loop controller is essentially used to minimize or eliminate the error in an output. The proposed digital control algorithm's flow chart and the control loop for the DAB-IBDC are presented in Figs. 3.8 and 3.9. The four variables—output voltage (V_{out}), input voltage (V_{in}), input current (I_{in}), and output current (I_{out})—of the DAB-IBDC are given to the DSP as processing variables through voltage and current sensors. LV25-P (LEM Components, Geneva, Switzerland) and LA25-P (LEM Components, Geneva, Switzerland) were used as voltage and current sensors. These sensors sense the input/output (I/O) voltages and I/O currents of the DAB-IBDC and send the corresponding low-voltage signal to the DSP. The DSP controller operates the MOSFET switches in a higher switching frequency. It generates high-

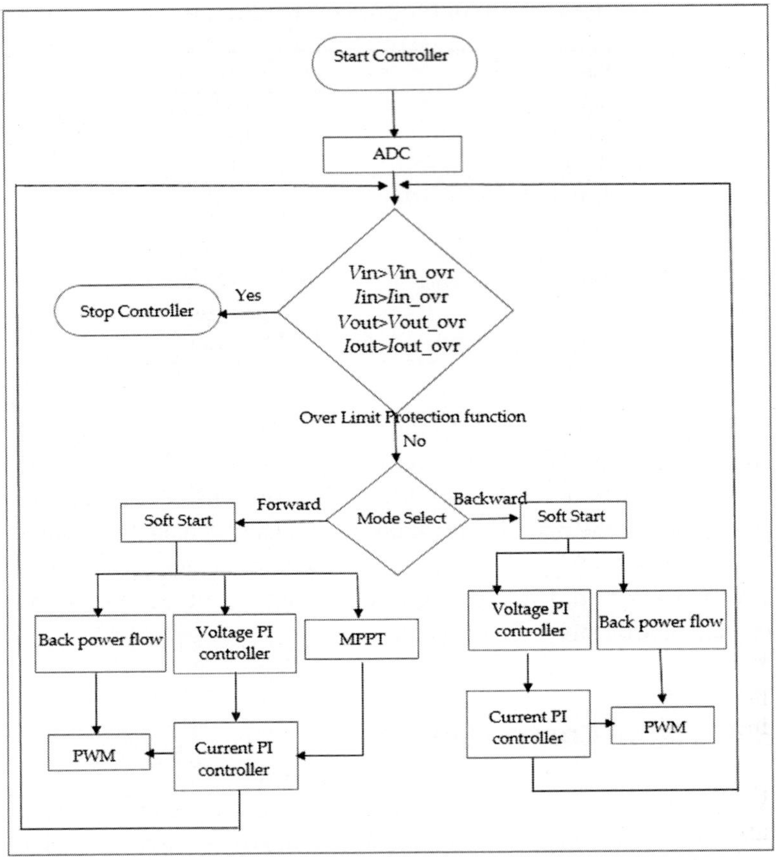

Figure 3.8 Blended SPS-ESPS digital control algorithm flow chart.

frequency, 50% duty cycle PWM signals for triggering MOSFET switches located on bridges on both sides. Along with this, the DSP generates the inner and outer phase angle on the basis of the proposed control algorithm, which controls the DAB-IBDC toward achieving safe charging, MPP, and high-efficiency power conversion. In DSP control, proportional integral (PI) controller has been used for achieving zero steady-state error in an output by finding the proper phase angle:

$$\text{PI controller} = K_{\text{p}}\text{e}\,(t) + \frac{1}{K_{\text{i}}}\int_0^t \text{e}(\tau)d\tau \tag{3.17}$$

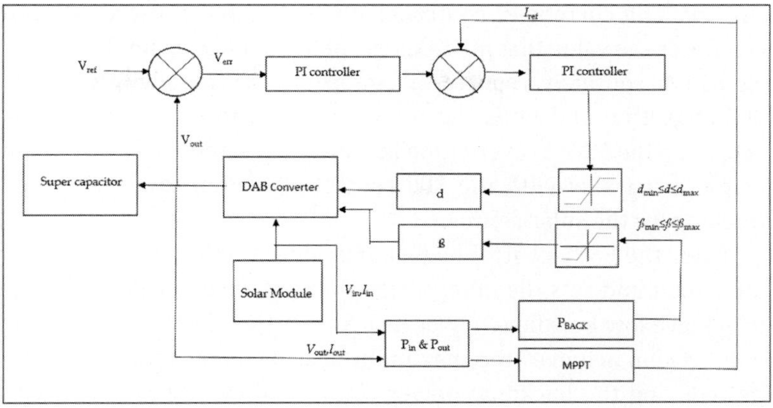

Figure 3.9 Blended SPS-ESPS digital control algorithm control loop.

The digital PI controller is given by:

$$\begin{aligned}\text{PI\_out} &= \text{Pre\_PI\_out} + \text{Err} \times (K_p + 0.5 * K_i * T_s) \\ &\quad + \text{Pre\_Err} \times (0.5 * K_i \times T_s - K_p), \end{aligned} \tag{3.18}$$

where Err = ref variable, Pre_Err = previous Err value, Pre_PI_out = previous PI_out value, K_p = proportional controller gain, and K_i = integral controller gain.

The proposed blended SPS-ESPS digital control algorithm initially performs the control process started with the SPS control algorithm. The controller generates an error signal based on the reference and output voltage. The voltage PI controller processes that error signal and finds the value for the outer phase shift (d). Thus, while starting, the DAB-IBDC inevitably sets the phase shift between the two bridges' square waves V_{ab} and V_{cd} using the SPS control algorithm. This helps in achieving safe charging current (Eq. 3.3) for the supercapacitors by controlling the inductor current (I_{L}). Meanwhile, the MPPT algorithm runs parallelly and continuously monitors the input power. At every cycle, it frequently varies the current reference value (I_{ref}) and subsequently monitors the corresponding changes in the input power. In 6 to 10 DSP timing cycles, the algorithm finds the better reference value where the maximum input power can be obtained. The error signal is generated by comparing this reference value with the supercapacitor charging current. The current PI controller processes that error signal and generates the value that modifies the phase shift '*d*' again. The MPPT algorithm routinely repeats the same process parallelly with the SPS algorithm and finds the reference value toward continuously achieving the MPP at every time instance. Thus, the MPPT algorithm works in parallel with the SPS control algorithm to achieve MPP tracking for the solar source.

Later the ESPS control begins to analyze the backflow power of the circuit and sets the inner phase shift (β) for the primary bridge to achieve low backflow power in a circuit (Eqs. 3.14 and 3.15). The ESPS region achieves the maximum possible conversion efficiency by reducing the backflow power. The backflow power of a circuit depends on the voltage conversion ratio '*k*' ($k = V_{\text{in}}/n \times V_{\text{out}}$). For large input to output voltage variations, '*k*' will be greater than 1. According to Shi et al.,[21] a comparison of ESPS with the SPS control algorithm showed that SPS control DAB converter backflow power is low for a particular phase-shift region, that is, 0.41 to 0.5 for $k > 2$. In the rest of that region, the ESPS control algorithm achieves higher efficiency than SPS by achieving low backflow power.

According to the mentioned DAB-IBDC design, '*k*' is:

$$k = \frac{V_{\text{in}}}{nV_{\text{out}}} = \frac{200 \sim 222}{0.75 * 48} \simeq 6,$$

where $k =$ voltage transfer ratio.

From Eqs. 3.6 and 3.15, it is noted that the backflow power (P_{bf}) is directly proportional to the voltage transfer ratio '*k*', so the backflow power for this DAB-IBDC design will also be the maximum as it has the maximum *k* value. Therefore, to increase the power conversion efficiency in such a large input to output voltage variation, this chapter proposes a combination of SPS and ESPS control algorithm. This works in such a way that for a particular outer-phase-shift '*d*' region, it will follow the SPS control algorithm and for the remainder of the outer-phase-shift '*d*' region, it will obey the ESPS algorithm. Thus, the proposed control algorithm operates in both SPS and ESPS modes for achieving the maximum possible power conversion efficiency in a large voltage conversion ratio. When conversion is started, the proposed control algorithm initially operates the converter with a soft start algorithm to reduce the inrushing current and overshoot voltage in a circuit. Next, the

SPS control and MPPT algorithm are started. Parallel working of the SPS control and the MPPT algorithm helps in enhancing the safety charging process for the supercapacitor bank while achieving the solar MPP. After that the ESPS algorithm is initiated and starts to analyze the backflow power in a circuit. On the basis of this analysis, the inner phase shift of the primary bridge is adjusted or maintained at zero toward achieving the maximum possible power conversion efficiency. The proposed blended SPS-ESPS digital control 2 kW DAB-IBDC performance was analyzed experimentally with a comparison study of a traditional SPS digital control DAB-IBDC. Both experiment results are discussed below. The proposed control algorithm achieved better conversion efficiency than the tradition control algorithm in a large voltage conversion ratio, creating a renewable energy storage system.

3.5 Experiment Results and Discussion

The power conversion performances of both SPS and blended SPS-ESPS control DAB-IBDC discussed below are based on the experimental results. The experiments were conducted in a stand-alone renewable energy storage system with a combination of 2 kW solar modules and a 153 Wh supercapacitor bank. The experiments were conducted during different time stages toward achieving different input power and voltages by the solar module. The DAB-IBDC output waveforms are discussed for the various input power values, 2000 W, 1900 W, 1700 W, and 1400 W, and the circuit performance is analyzed using the experimental results. The SPS control DAB-IBDC conversion waveform, charging time, and efficiency are discussed below.

An SPS control is a traditional control technology for a DAB-IBDC. Figure 3.10 shows that the SPS controls the DAB-IBDC conversion waveforms for different input powers. The HV (V_{ab}) and LV (V_{cd}) side voltage, inductor current, and output charging voltage are shown for each input power. It shows that DSP automatically varies the phase shift (0.47 to 0.245) between V_{ab} and V_{cd} to achieve a constant output voltage and limiting current based on the SPS control algorithm. The solar source achieved an input voltage from

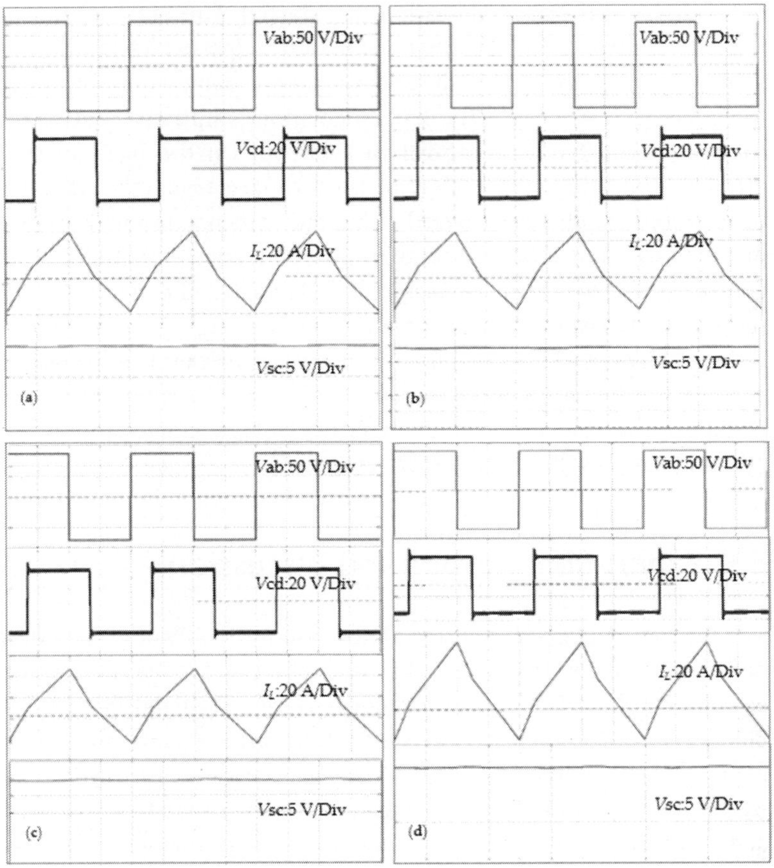

Figure 3.10 Experimental results of the SPS control DAB-IBDC at different input power values: (a) 2000 W, (b) 1900 W, (c) 1700 W, and (d) 1400 W. Timescale: 5 ms/Div.

a to d of 221.7 V, 216.3 V, 211.1 V, and 207.7 V, respectively. The SPS digital control algorithm changed only the outer phase shift to control the inductor current and obtain a constant DC output voltage.

Figure 3.11 presents the supercapacitor charging waveform for SPS control DAB-IBDC. Figure 3.11a shows the time-versus-voltage characteristic waveforms, and Fig. 3.11b shows the time-versus-current characteristic waveforms of the converter. It takes 6 min.

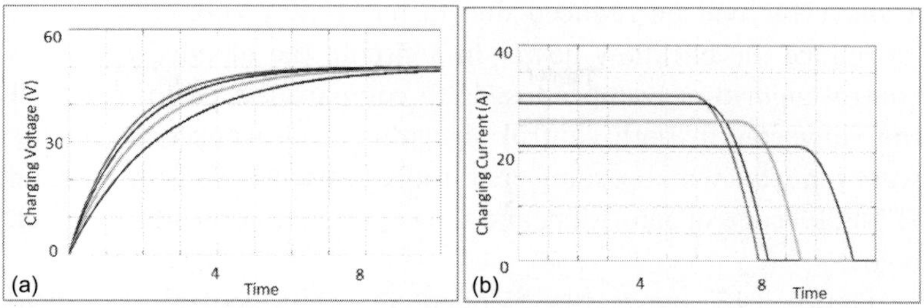

Figure 3.11 Supercapacitor charging waveform by an SPS control DAB-IBDC.

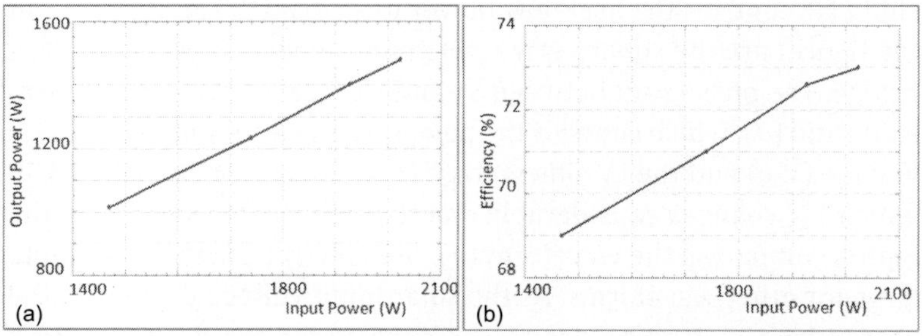

Figure 3.12 SPS control DAB-IBDC power conversion efficiency for different values of input solar power.

26 s, 6 min. 58 s, 7 min. 43 s, and 9 min. 23 s to charge a 158 Wh supercapacitor bank with input power of 2000 W, 1900 W, 1700 W, and 1400 W, respectively. On the other hand, according to the theoretical calculation, input energy of 2000 W, 1900 W, 1700 W, and 1400 W can charge a 158 Wh supercapacitor bank in 4 min. 45 s, 5 min. 1 s, 5 min. 35 s, and 6 min. 47 s, respectively. Owing to the power losses in a circuit SPS control algorithm achieves less efficiency for the given prototype. The efficiency of the converter can be calculated from the difference between the theoretical and experimental values, which are plotted in Fig. 3.12.

Figure 3.12a,b shows the achieved output power and power conversion efficiency of the SPS control DAB-IBDC. The SPS control topology achieved power conversion efficiency of 73%, 72.6%, 71%, and 69% for input solar power of 2000 W, 1900 W, 1700 W, and 1400 W, respectively. As discussed above, the efficiency of a converter will be reduced due to backflow power in a circuit. To reduce the backflow power in a circuit, the blended SPS-ESPS control algorithm was proposed. To compare the performance of the converter in both control algorithms, the same experiments were repeated with the same operating conditions using the blended SPS-ESPS control algorithm, and the experimental results are given below.

Figure 3.13 shows the blended SPS-ESPS control DAB-IBDC circuit waveforms. The two square wave signals S1 and S4 are the switching signals for the two inner branches of the primary bridge. The phase shift between these two branches, called the inner phase shift (β), controls the backflow power in a circuit. S5 and S7 (S5,7 in the figure) are the square wave switching signals for the secondary bridge. The phase shift between S1 and S5 is called the outer-phase-shift ratio (d), which controls the power flow in a circuit. The HV side voltage (V_{ab}) and the LV side voltage (V_{cd}) are shown in Fig. 3.13, in which V_{ab} changes considerably due to the inner phase shift. In the figure, comparing the waveform (a), the inner phase shift was quite large for other waveforms. As the solar input voltage decreases, the outer phase shift increases to compensate for the desired output DC voltage. Therefore, the backflow power in a circuit increases and subsequently the conversion output power of the circuit is decreased. It shows the proposed algorithm to reduce the backflow power by increasing the inner phase shift. Because the additional phase shift voltage difference across the inductor has six levels of changes, it resembles the changes in the inductor current.

Figure 3.14 shows the AC link current waveforms generated by the two active bridges. Figure 3.14a,b shows the LV- and HV-side device current waveforms. From the figure, introducing an inner phase shift in an SPS reduces the switching reverse current or the backflow power. From the measured results, the blended SPS-ESPS control converter operates with a maximum conversion efficiency of 84%.

Figure 3.15 presents a power conversion efficiency comparison plot for SPS and proposed control algorithm; Fig. 3.15a shows the power conversion efficiency and Fig. 3.15b shows the achieved output power respective to various values of input solar power. The blended SPS-ESPS topology achieved power conversion efficiency of 75%, 79%, 84%, and 81% for input solar power of 2000 W, 1900 W,

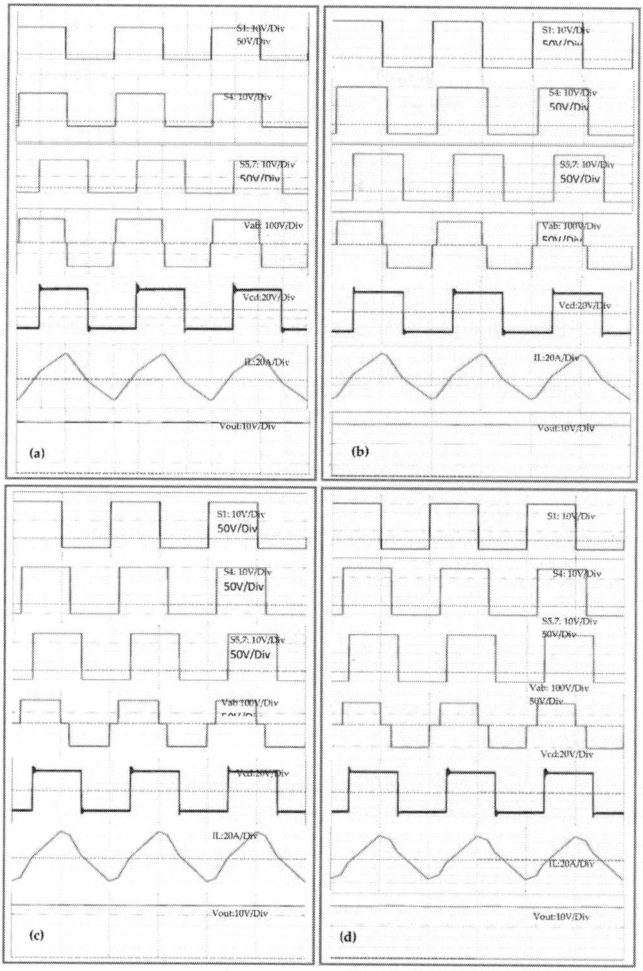

Figure 3.13 Blended SPS-ESPS control DAB-IBDC conversion waveform for different values of input power: (a) 2000 W, (b) 1900 W, (c) 1700 W, and (d) 1400 W. Timescale 5 ms/Div.

1700 W, and 1400 W, respectively. Compared to the SPS control algorithm, it increases the power conversion efficiency by 2%–13% and takes 6 min. 14 s, 6 min. 15 s, 6 min. 30 s, and 7 min. 59 s to charge a 158 Wh supercapacitor bank with input power of 2000 W, 1900 W, 1700 W, and 1400 W, respectively.

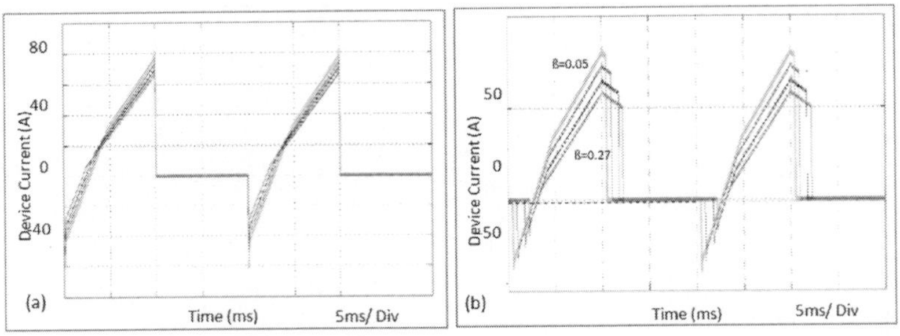

Figure 3.14 Experimental results of device current waveform of blended SPS-ESPS control DAB-IBDC. (a) Primary bridge switching current and (b) secondary bridge switching current.

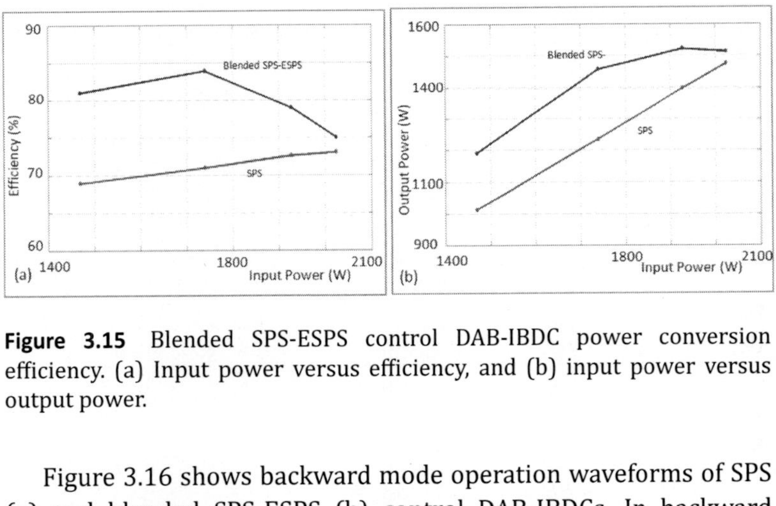

Figure 3.15 Blended SPS-ESPS control DAB-IBDC power conversion efficiency. (a) Input power versus efficiency, and (b) input power versus output power.

Figure 3.16 shows backward mode operation waveforms of SPS (a) and blended SPS-ESPS (b) control DAB-IBDCs. In backward mode, a high-power resistive load (25 Ω) was used on the primary side of the converter circuit, as shown in the block diagram in Fig. 3.1, for discharging the supercapacitor bank. In forward mode, when the supercapacitor was fully charged, the controller automatically switched from forward mode to backward mode and enabled reverse-side power flow in the circuit. In the experiment, a 25 Ω, 2500 W resistive load was connected on the primary side to discharge a 158 Wh supercapacitor bank. The supercapacitor bank output voltage was 48 V, which was converted to 220 V

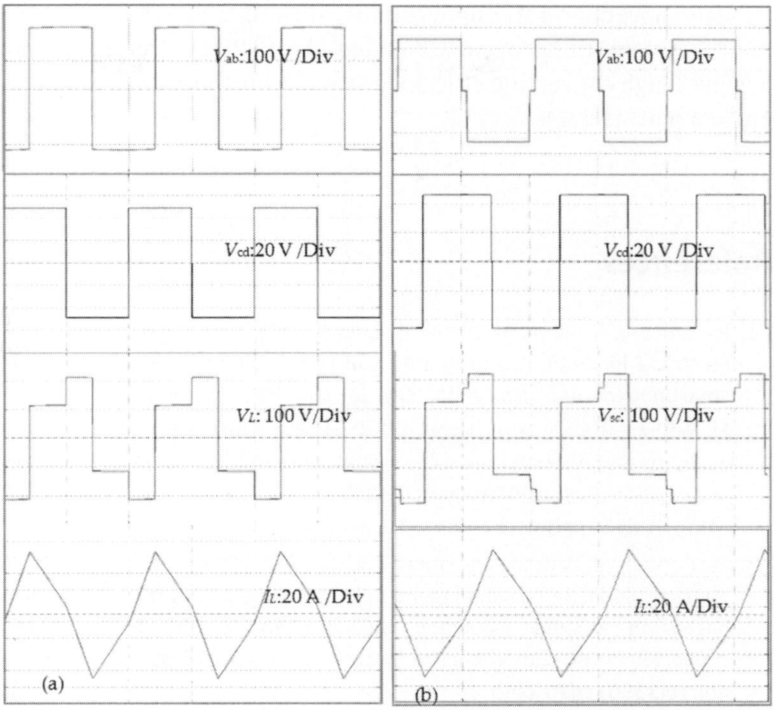

Figure 3.16 Backward mode operation waveforms of (a) SPS control and (b) blended SPS-ESPS control DAB-IBDCs. Timescale 10 ms/Div.

DC in backward mode and applied across the load of 25 Ω. An approximately 9 A current was passed through the load.

3.6 Conclusions

The blended SPS-ESPS digital control algorithm was proposed over the SPS control algorithm for a DAB-IBDC to achieve high power conversion efficiency, MPP, and enhanced fast charging-discharging in a high-power-density energy storage device. In the laboratory, using MOSFET switches, a 2 kW, 50 kHz DAB-IBDC was developed and the SPS and blended SPS-ESPS control algorithm were verified experimentally. The results showed that in large-voltage-conversion-ratio applications, such as a stand-alone solar power system, the proposed blended SPS-ESPS control algorithm achieves high conversion efficiency by reducing the backflow power and current stress in a circuit.

References

1. R. Romano, P. Siano, M. Acone, V. Loia (2017). Combined operation of electrical loads, Air conditioning and photovoltaic battery systems in smart houses, *Appl. Sci.*, **7**, pp. 525, doi:10.3390/app7050525.
2. M. A. Khan, B. Ko, E. Alois Nyari, S. E. Park, H.-J. Kim (2017). Performance evaluation of photovoltaic solar system with different cooling methods and a bi-reflector PV system (BRPVS): an experimental study and comparative analysis, *Energies*, **10**, p. 826.
3. M. Daowd, M. Antoine, N. Omar, P. Lataire, P. van den Bossche, J. van Mierlo (2014). Battery management system—balancing modularization based on a single switched capacitor and bi-directional DC/DC converter with the auxiliary battery, *Energies*, **7**, pp. 2897–2937, doi:10.3390/en7052897.
4. C. Xiang, Y. Wang, S. Hu, W. Wang (2014). A new topology and control strategy for a hybrid battery-ultracapacitor energy storage system, *Energies*, **7**, pp. 2874–2896, doi:10.3390/e,7052874.
5. S. W. Sheldon (2013). Electric and plug-in hybrid electric vehicle drive train topologies, in *Energy Management Strategies for Electric and Plug-in Hybrid Electric Vehicle*, Springer, New York, NY, USA, pp. 7–14.
6. B. Balamuralitharan, S. N. Karthick, S. K. Balasingam, K. V. Hemalatha, S. Selvam, J. Anandha Raj, K. Prabakar, Y. Jun, H.-J. Kim (2017). Hybrid reduced graphene oxide/manganese diselenide cubes: a new electrode material for supercapacitors, *Energy Technol.*, **5**, pp. 1953–1962, doi: 10.1002/ente.201700097.
7. C. M. Lai, Y. C. Lin, D. S. Lee (2015). Study and implementation of a two-phase interleaved bidirectional DC/DC converter for vehicle and DC-microgrid systems, *Energies*, **8**, pp. 9969–9991.
8. C. T. Tsai, W.-M. Chen (2016). Buck converter with soft switching cells for PV panel applications, *Energies*, **9**, pp. 148, doi:10.3390/en9030148.

9. H. Patel, V. Agarwal (2009). MPPT scheme for a PV-fed single-phase single-stage grid-connected inverter operating in CCM with only one current sensor, *IEEE Trans. Energy Convers.*, **24**, pp. 256–263.
10. Y. P. Hsieh, J.-F. Chen, L.-S. Yang, C.-Y. Wu, W.-S. Liu (2014). High-conversion-ratio bidirectional DC-DC converter with coupled inductor, *IEEE Trans. Ind. Electron.*, **61**, pp. 210–222.
11. R. W. A. A. De Doncker, D. M. Divan, M. H. Kheraluwala (1991). A three-phase soft-switched high-power-density dc/dc converter for high-power applications, *IEEE Trans. Ind. Appl.*, **27**, pp. 63–73.
12. J. Biela, M. Schweizer, S. Waffler, W. J. Kolar (2011). SiC versus Si—evaluation of potentials for performance improvement of inverter and DC-DC converter systems by SiC power semiconductors, *IEEE Trans. Ind. Electron.*, **58**, pp. 2872–2882.
13. A. R. Alonso, J. Sebastian, D. G. Lamar, M. M. Hemando, A. Vazquez (2010). An overall study of a dual active bridge for bidirectional DC/DC conversion, in *Proceedings of the 2010 IEEE Energy Conversion Congress and Exposition (ECCE)*, Atlanta, GA, USA, doi:10.1109/ECCE.2010.5617847.
14. B. Zhao, Q. Song, W. Liu, Y. Sun (2014). Overview of dual-active-bridge isolated bidirectional DC-DC converter for high-frequency-link power conversion system, *IEEE Trans. Power Electron.*, **29**, pp. 4091–4106.
15. R. T. Naayagi, J. Andrew, R. Shuttleworth (2012). High-power bidirectional DC-DC converter for aerospace applications, *IEEE Trans. Power Electron.*, **27**, pp. 4366–4379.
16. D.-K. Jeong, M.-H. Ryu, H.-G. Kim, H.-J. Kim (2014). Optimized design of bi-directional dual active bridge converter for low voltage battery charger, *J. Power Electron.*, **14**, doi:10.6113/JPE.2014.14.3.468.
17. S. Inoue, H. Akagi (2007). A bidirectional isolated dc-dc converter as a core circuit of the next-generation medium-voltage power conversion system, *IEEE Trans. Power Electron.*, **22**, pp. 535–542.
18. H. Zhou, A. M. Khambadkone (2009). Hybrid modulation for dual active bridge bi-directional converter with extended power range for ultracapacitor application, *IEEE Trans. Ind. Appl.*, **45**, pp. 1434–1442.
19. R. T. Naayagi, A. J. Forsyth, R. Shuttleworth (2015). Performance analysis of extended phase shift control of DAB DC-DC converter for aerospace energy storage system, in *Proceedings of the 2015 IEEE 11th International Conference on Power Electronics and Drive Systems (PEDS)*, Sydney, Australia.

20. B. Zhao, Q. Yu, W. Sun (2012). Extended-phase shift control of isolated bidirectional dc-dc converter for power distribution in microgrid, *IEEE Trans. Power Electron.*, **27**, pp. 4667–4680.
21. X. Shi, J. Jiang, X. Guo (2013). An efficiency-optimized isolated bidirectional dc-dc converter with extended power range for energy storage systems in microgrids, *Energies*, **6**, pp. 27–44, doi:10.3390/en6010027.

Chapter 4

Overview of Transformerless Inverter Structures for Grid-Connected PV Systems

Renewable energy (RE) sources are a very good solution to providing alternative energy to overcome the global energy issues. Further, the use of RE sources has improved during the last decade because of the development in grid integration technologies. However, there is a significant variation among grid-connected PV systems, from a few hundred watts (small-scale DC modules) to hundreds of megawatts (large scale). A large portion of residential applications consists of PV systems, and this will increase abruptly in the near future. This chapter comprehensively reviews the development and control of single-phase grid-connected (SPG-C) PV systems. A brief introduction of RE resources, along with their updated status, is provided in Section 4.1. Section 4.2 discusses power converter technology for PV systems. Half-bridge (H-bridge)-based inverter structures are detailed in Section 4.3. Section 4.4 summarizes FB-derived inverter topologies. NPC-derived inverter structures are elaborated in Section 4.5. NPC-derived inverter technologies, typical PV inverter structures, and generic control

Solar Power and Energy Storage Systems
Hee-Je Kim

ISBN 978-981-4800-13-6 (Hardcover), 978-0-429-45877-4 (eBook)
www.panstanford.com

structure for SPG-C systems are explained in Section 4.6, Section 4.7, and Section 4.8, respectively.

4.1 Introduction

According to long-term national strategies, the resources of conventional power generations are unsustainable. In the last decade, extensive installation of renewable energy resources (RERs), that is, hydropower, wind energy, solar photovoltaic (PV) energy, geothermal energy, ocean energy, biomass energy, tidal energy, and thermoelectric energy, has been carried out for grid interonnection.[1,2] Over the past several years, solar PV energy installation is booming at a repaid rate, and in some countries, it plays a vital role in electricity generation.[3] For instance, solar PV systems fulfilled approximately 7.9% of the total electricity demand in 2014 in Italy. By the end of 2014, its installed capacity reached 38.2 GW (mostly for residential purposes).[3–5] The worldwide solar PV generation capacity for the last 15 years is illustrated in Fig. 4.1. For desirable PV integration in the future mixed-power-grid and single-phase grid-connected (SPG-C) PV systems are discussed and described in this chapter.

The power electronics technology plays a vital role in the integration of RERs, including solar PVs, into the grid.[6,7] Power semiconductor devices are associated with advancement.[8] Therefore, the power electronics part is responsible for efficient and reliable power conversion from the inexhaustible, pollution-free, and clean solar PV energy. That's why, a large number of PV power converters for grids are advanced and commercialized.[9–15]

4.2 Power Converter Technology for PV Systems

In terms of progressing technologies, to organize and transfer PV power to a grid there are mainly five configuration concepts[2,9,16,17] available, as presented in Fig. 4.2. According to the power rating and the output voltage of the PV panels, each configuration consists of a series of paralleled strings or PV panels, followed and configured

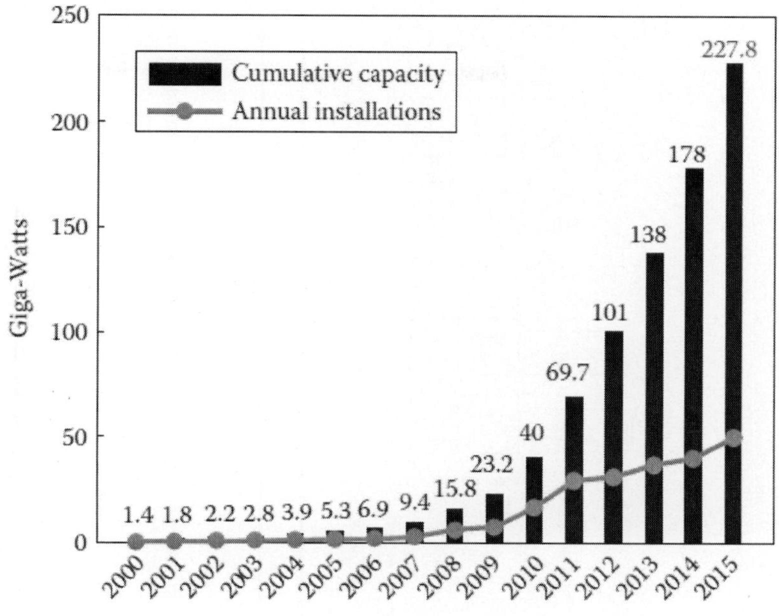

Figure 4.1 Global cumulative PV capacity for 15 years.[5]

by DC–DC converters and DC–AC inverters (power electronics converters).

Typically, power converters are classified into string inverters, multistring inverters, central inverters, and module-level (AC module and DC module) inverters.[9,10] For solar power farms/plants as three-phase systems, central and multistring converters are widely utilized.[18–20] Comparatively, in residential applications, as single-phase systems, string and module converters are intensively adopted.[21,22] Though the configuration of power converters is different, they have the same major functions, including islanding detection and protection, reactive power control, grid code compliance, synchronization, power transfer and DC–AC conversion, and PV power maximization.[7–22] Advanced and intelligent controls are required for effective incorporation of these functionalities and to fulfill customized demands. Additionally, PV integration can be enhanced by forecasting, monitoring, and communication technology.[18,21]

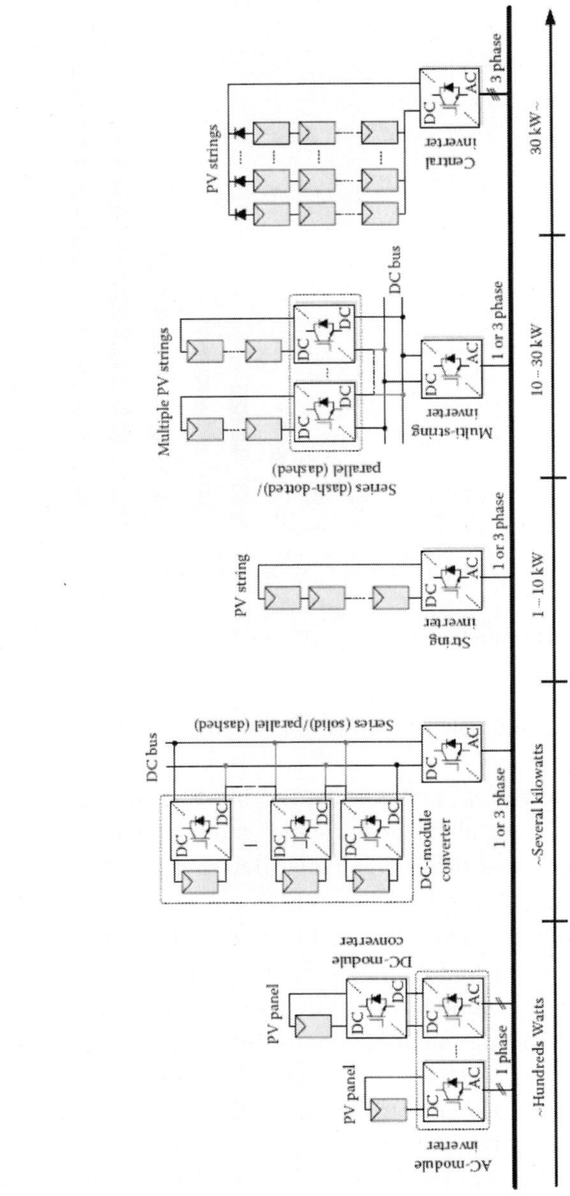

Figure 4.2 (a) residential/small systems, (b) residential/small systems, (c) residential, (d) residential /commercial, and (e) utility-scale/commercial PV applications.[24]

As presented in Fig. 4.2, the most common solutions used in single-phase PV applications are string inverters, multistring inverters, and the module concept, where an important issue of concern is galvanic isolation. Conventionally, isolation is provided by an isolation transformer either by a low-frequency transformer at the grid or by a high-frequency transformer between the power electronics converters, as presented in Fig. 4.2a and 4.2b. With an overall efficiency of 93%–95%, both above-mentioned grid-connected technologies are commercially available, contributed mainly by bulky transformers.[23] A large number of transformerless PV inverters have been developed,[9,11,23] and there is daily improvement in the overall efficiency.

Recently, many manufacturers of transformerless PV inverters have entered the market, for example, REFU, Danfoss, Ingeteam, Conergy, Sunways, and SMA, offering maximum efficiency of up to 98% and high European efficiency (>97%). The topology development for transformerless inverter is based on the following main converter families:

(a) H-bridge or full-bridge
(b) Neutral point clamped (NPC)

Based on these main families, most relevant derived transformerless topologies are discussed and explained in this chapter. In some structures a boost DC–DC converter is required. That's why the level of diversity is high. Boost converters are not described in detail as they are well known. At the end of this chapter some classical combined boost converter topologies are described.

4.3 H-Bridge-Based Inverter Structures

In 1965, W. Mcmurray first developed the full-bridge (FB) or H-bridge inverter family.[25] In the power electronics converter technology development, the FB has been an important development. Force-commuted semiconductor devices, also called thyristors, were first effectively utilized by this structure. This topology can be implemented with one switching leg (in the H-bridge form) or with two switching legs (in the FB form) and can be used for both

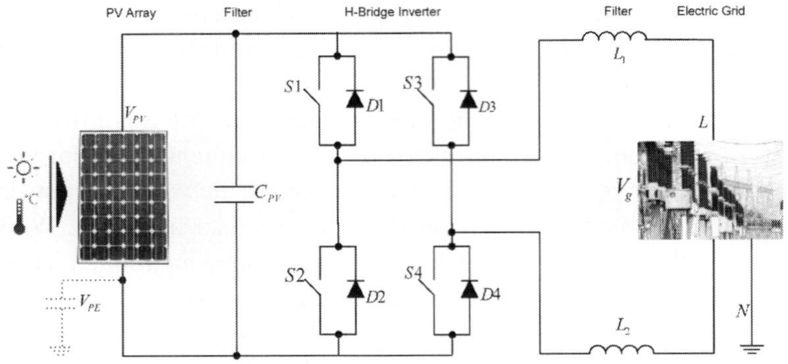

Figure 4.3 Basic structure of an FB inverter.

DC–AC and DC–DC conversion; it's also a versatile topology. The basic structure of the FB is presented in Fig. 4.3.

4.3.1 Modulation Strategies

The basic modulation strategies for an inverter structure are:

- Bipolar (BP) modulation
- Unipolar (UP) modulation
- Hybrid modulation

4.3.1.1 Bipolar modulation

In this modulation the diagonal switches are turned on as S1 with S4 and S2 with S3, respectively. For the positive and negative output currents, the AC voltage can be generated as presented in Figs. 4.4a and 4.4b, respectively. This converter is based on the following features:

- The diagonal switches are synchronously switched on, that is, S1 with S4 and S2 with S3 at a high frequency.
- A voltage state with zero output is not possible.

Aspects of significance

The electric magnetic interference (EMI) and leakage current are very low as V_{PE} has no switching frequency component while only a grid frequency component is present.

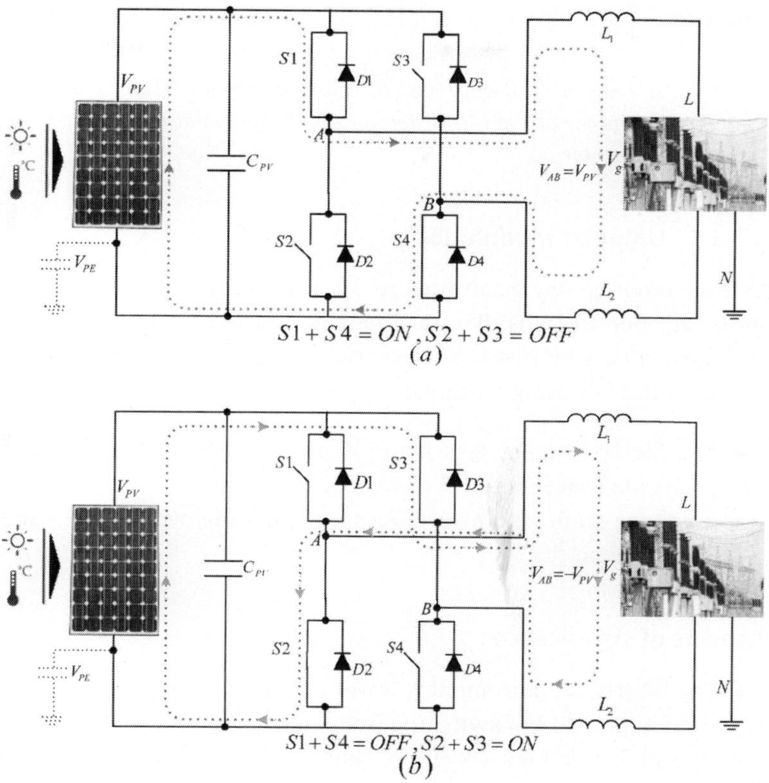

Figure 4.4 BP modulation scheme for an FB in the case of (a) a positive cycle and (b) a negative cycle.

Aspects of insignificance

- The efficiency is low, up to 95%, because of simultaneous switching of the two switches at every period the core losses in the output filter are higher and reactive power exchange between $L_{1\leftrightarrow 2}$ and C_{PV} during freewheeling.
- The filtering requirement is higher as the switching ripple in the current equals 1 × switching frequency.
- Due to BP voltage variation, that is, $(+V_{PV} \rightarrow -V_{PV} \rightarrow +V_{PV})$ the core losses are higher.

Analysis

Due to reduced efficiency the BP modulation–based FB is not appropriate for use in a transformerless PV system despite its low leakage current.

4.3.1.2 Unipolar modulation

In this modulation, according to its own reference each leg is switched. For the positive and negative output currents, the AC voltage can be generated as presented in Fig. 4.5. This converter is based on the following features:

- At a high frequency legs A and B are switched with a mirrored sinusoidal reference.
- A voltage state with a zero output is possible, when S1, S3 and S2, S4 are on.

Aspects of significance

- The filtering requirement is lower as the switching ripple in the current equals 1 × switching frequency.
- Due to UP voltage variation, that is, ($0 \rightarrow +V_{PV} \rightarrow 0 \rightarrow -V_{PV} \rightarrow 0$), the core losses are lower.
- Because of reduced losses during the zero-voltage states its efficiency is higher, up to 98%.

Aspects of insignificance

- The EMI and leakage current are very high as V_{PE} has no switching frequency component.

4.3.1.3 Analysis

Due to the high-frequency content of the V_{PE} the UP modulation–based FB is not appropriate for use in a transformerless PV system, although it has high efficiency and low filtering requirements.

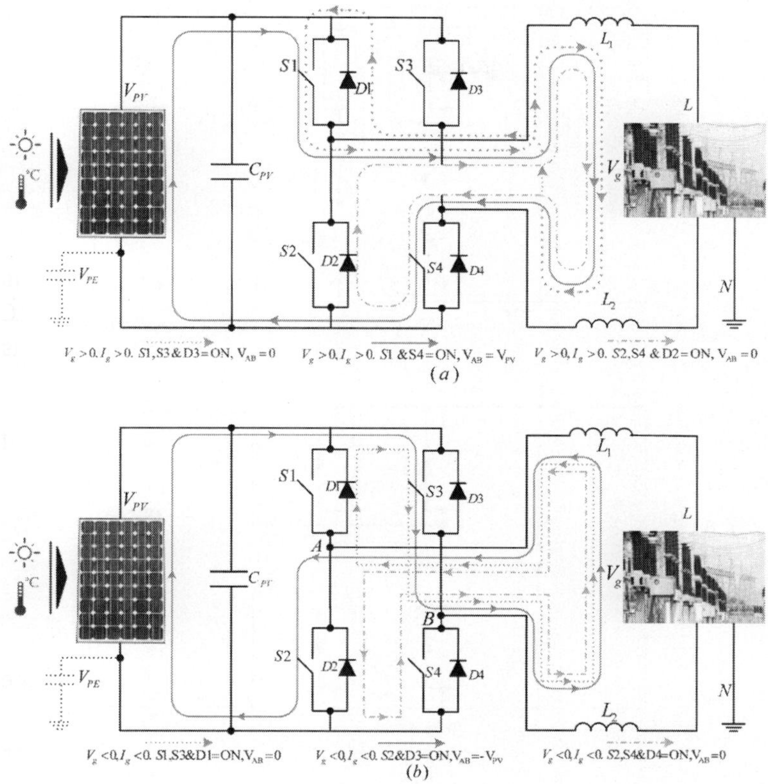

Figure 4.5 UP modulation scheme for an FB in the case of (a) a positive cycle and (b) a negative cycle.

4.3.1.4 Hybrid modulation

According to hybrid modulation one leg is switched at a higher frequency and the other leg is switched at the grid frequency.[25] For the positive and negative output currents, the AC voltage can be generated as described in Fig. 4.6. This converter is centered on the subsequent features.

- At a high pulse-width-modulation (PWM) frequency leg A is switched on while leg B is switched at grid low frequency.
- A voltage state with two zero outputs is possible, when S1, S2 and S3, S4 are on.

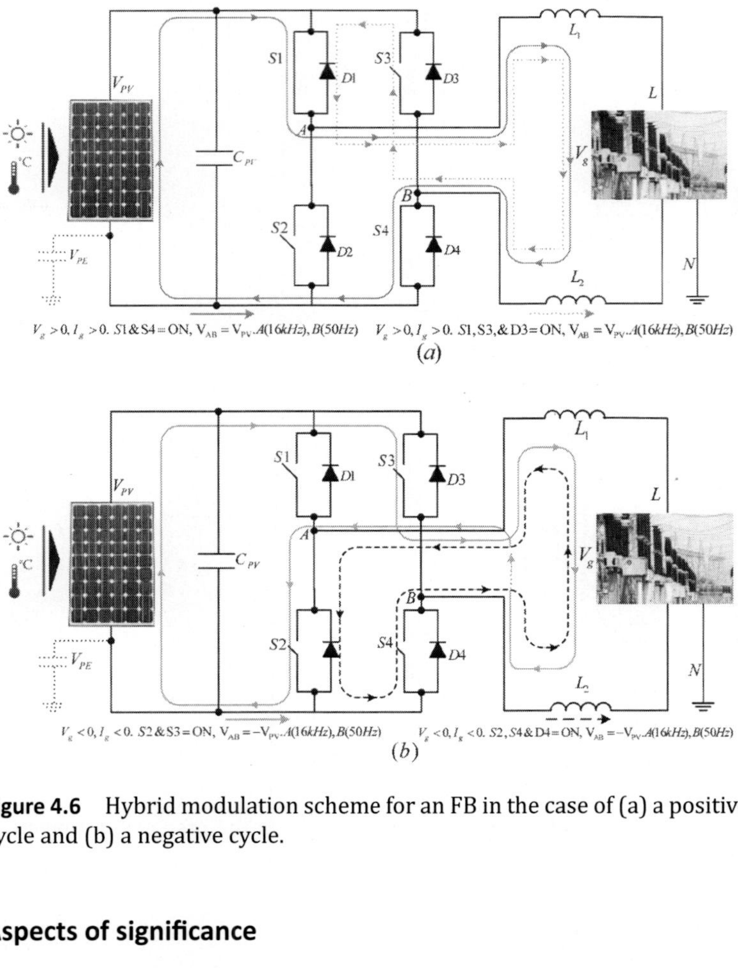

Figure 4.6 Hybrid modulation scheme for an FB in the case of (a) a positive cycle and (b) a negative cycle.

Aspects of significance

- Due to UP voltage variation, that is, $(0 \rightarrow +V_{\mathrm{PV}} \rightarrow 0 \rightarrow -V_{\mathrm{PV}} \rightarrow 0)$, the core losses are lower.
- Because of no reactive power exchange between C_{PV} and $L_{1(2)}$ during zero voltage and lower frequency of switching in one leg, its efficiency is higher, up to 98%.

Aspects of insignificance

- The EMI filtering requirement and the leakage current peaks are high as V_{PE} has square wave variation at the grid frequency.

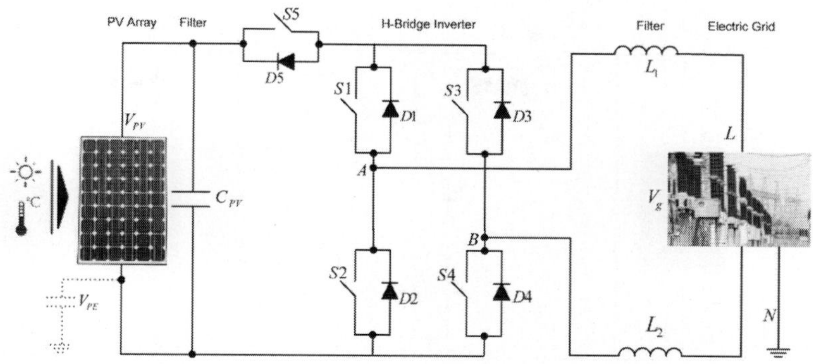

Figure 4.7 Schematic diagram of an H5 (SMA) inverter.

- The filtering requirement is higher (in the output no artificial frequency increase) as the switching ripple in the current equals 1 × switching frequency.
- For a two-quadrant operation only this modulation works.

Analysis

Due to the square wave variation of the V_{PE} the hybrid modulation–based FB is not appropriate for use in a transformerless PV system although it has high efficiency.

4.3.2 H5 Inverter (SMA)

A new inverter topology called H5 was patent by SMA in 2005.[26] As its name indicates, it is a modified H-bridge. In the positive bus of the DC link an extra switch—a fifth switch—is added, presented in Fig. 4.7. The extra switch has the following two vital functions:

- The efficiency is increased as no reactive power exchange occurs between C_{PV} and $L_{1(2)}$ during zero voltage.
- It eliminates the high-frequency content of V_{PE} by isolating the grid from the PV modules during the zero-voltage state.

Figure 4.8 presents the generation of AC currents for the positive and negative switching states. This converter is based on the succeeding features:

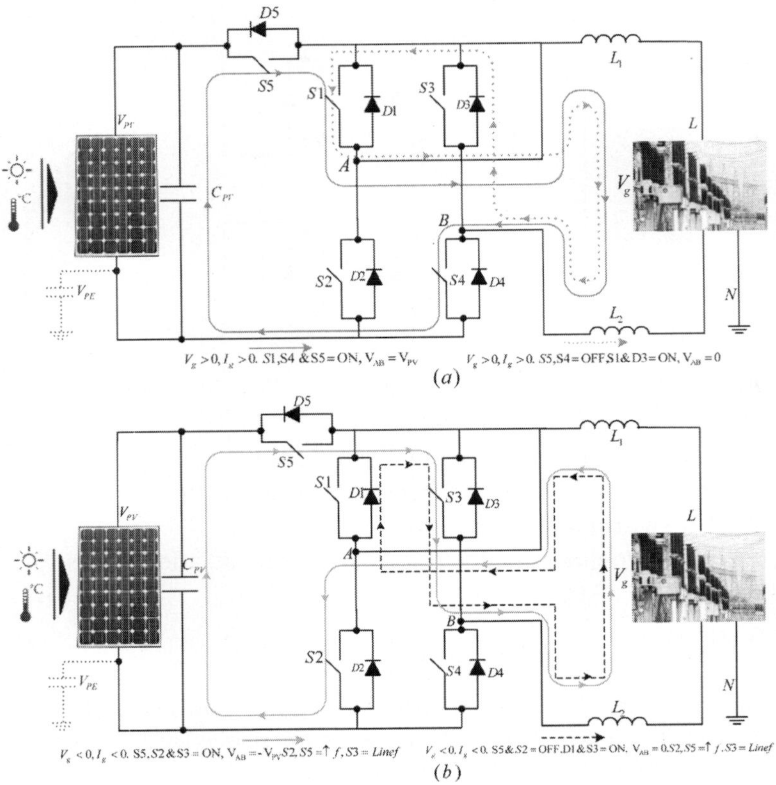

Figure 4.8 Switching states of H5 (SMA) in the case of (a) a positive cycle and (b) a negative cycle.

- Voltage states with two zero outputs are possible, that is, when S5 is off and when S4(S2) is on.
- S1 and S3 are switched at the grid frequency; and S2, S4, and S5 are switched at a high frequency.

Aspects of significance

- The core losses are lower due to UP voltage variation, that is, $(0 \rightarrow +V_{PV} \rightarrow 0 \rightarrow -V_{PV} \rightarrow 0)$.
- Because of no reactive power exchange between C_{PV} and $L_{1(2)}$ during zero voltage and lower frequency of switching in one leg, its efficiency is higher, up to 98%.

- The EMI filtering requirement and the leakage current peaks are lower as V_{PE} has only a grid frequency component.

Aspects of insignificance

- The conduction losses are higher as three switches are conducting during the active vector, but the overall high efficiency is not affected.
- An extra switch is added.

Analysis

*The advantageous features of an FB with a hybrid modulation technique is combined in the H5 topology. On utilization of the extra switch, the high-frequency component of V_{PE} is eliminated by isolating the grid from PV panels during the zero-voltage state. Due to the high efficiency, low filtering requirements, and low EMI, this topology is thus very appropriate for use in a transformerless PV system. Currently, in the series called SunnyBoy 4000/5000 TL, it is commercialized by SMA with a maximum efficiency of 98% (*Photon International*, October 2007) and European efficiency higher than 97.7%.*

4.3.3 HERIC Inverter (Sunways)

A new inverter topology called "highly efficient and reliable inverter concept" (HERIC) was patented by Sunways in 2006. This topology is also called a highly efficient and reliable inverter concept. As presented in Fig. 4.9, using two back-to-back connected insulated gate BP transistors (IGBTs), a bypass leg is added on the AC side.[27] The AC bypass is used to give the same vital functions as the extra—fifth—switch in the H5 topology, as follows:

- The efficiency is increased as no reactive power exchange occurs between C_{PV} and $L_{1(2)}$ during zero voltage.
- It eliminates the high-frequency content of V_{PE} by isolating the grid from the PV modules during the zero-voltage state.

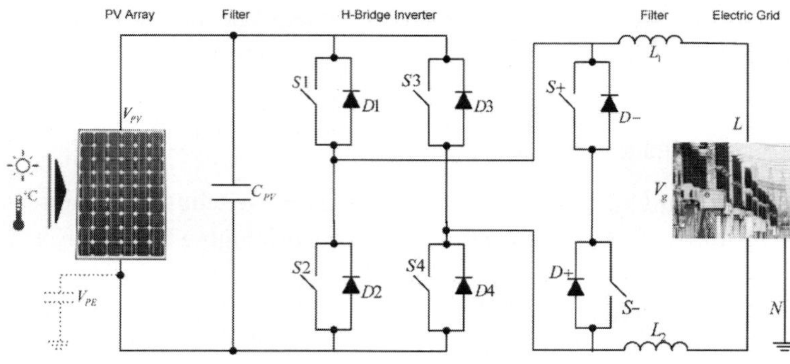

Figure 4.9 Schematic diagram of an HERIC (Sunways) inverter.

Figure 4.10 presents the generation of AC currents for the positive and negative switching states. This converter is established on the following features.

- Voltage states with two zero output values are possible, that is, when S+ and S– are on (when the bridge is in the off state).
- S+ (S–) is switched at the grid frequency, and S1–S4 and S2–S3 are switched at a high frequency.

Aspects of significance

- The core losses are lower due to UP voltage variation, that is, $(0 \rightarrow +V_{PV} \rightarrow 0 \rightarrow -V_{PV} \rightarrow 0)$.
- Because of no reactive power exchange between C_{PV} and $L_{1(2)}$ during zero voltage and lower frequency of switching in one leg its efficiency is higher, up to 97%.
- The EMI filtering requirement and the leakage current peaks are lower as V_{PE} has no switching frequency component and only a grid frequency component is present.

Aspects of insignificance

- Two extra switches have been added.

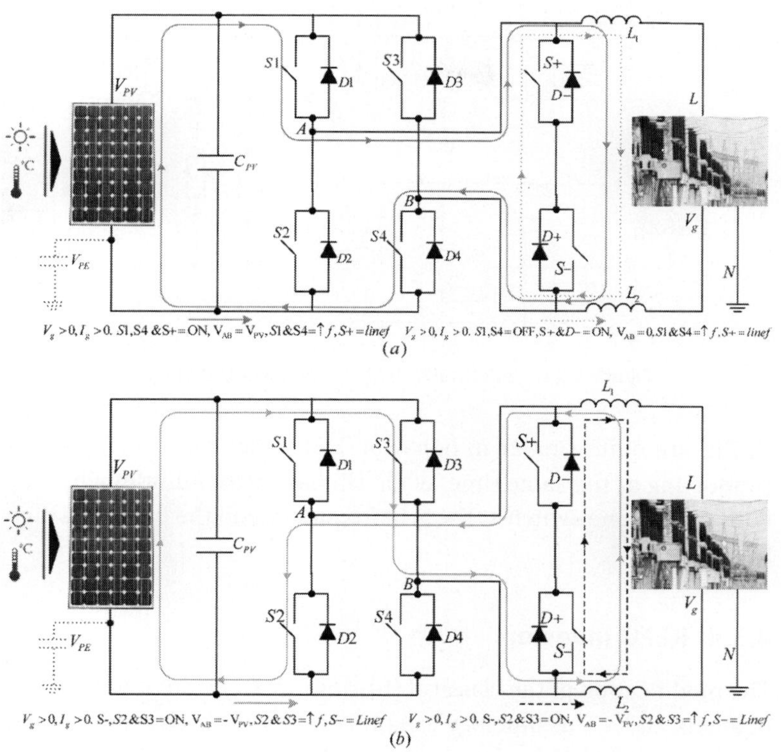

Figure 4.10 Switching states of HEREIC (Sunways) in the case of (a) a positive cycle and (b) a negative cycle.

Analysis

> *The HERIC topology increases the efficiency by adding the zero voltage with the help of an AC bypass to the performance of the FB with BP modulation.*
>
> *Due to the high efficiency, low filtering requirements, and low EMI, this topology is very appropriate for use in a transformerless PV system. Currently, in the series called Sunways in the AT series (2.7–5 kW) it is commercialized, with maximum efficiency of 95.6%* (Photon International, *July 2008) and European efficiency of 95%.*

As during the zero-voltage switching decoupling of the grid from the PV generator on the AC side and DC side occurs, both H5 and

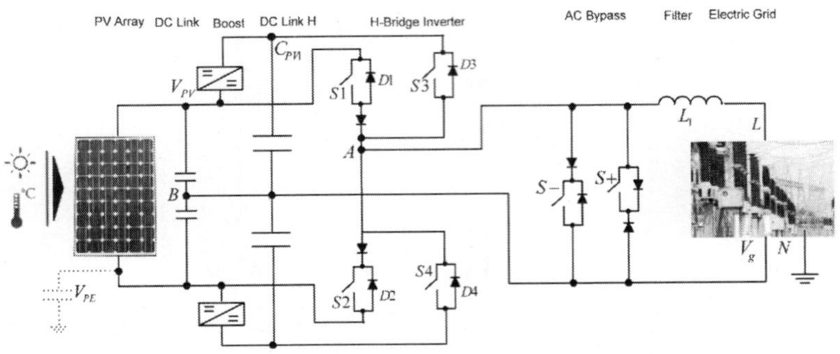

Figure 4.11 Schematic diagram of a REFU inverter.

HERIC are quite similar in behavior. HERIC has only two switches conducting at the same time, while H5 has three. Additionally, they both switch, one switch at the grid frequency and the other at a high frequency.

4.3.4 REFU Inverter

The modification of the classical H-bridge by REFU solar gave a new patent in 2007. This topology usually consists of a bypassable DC–DC converter and an H-bridge within the AC side bypass, as presented in Fig. 4.11.[28] The same two vital functions of HERIC topology are provided by the AC bypass.

- The efficiency is improved as no reactive power exchange happens between C_{PV} and $L_{1(2)}$ during zero voltage.
- It removes the high-frequency content of V_{PE} by detaching the grid from the PV modules during the zero-voltage state.

In comparison to HERIC the AC bypass is differently implemented, for example, the standard IGBT module–based unidirectional switches are used and a series diode with an IGBT is used to cancel the freewheeling path. In this topology, when the input DC voltage is lower than the grid voltage, only the boost converter is activated. Figure 4.12 presents the generation of AC currents for the positive and negative switching states. This converter is established on the following features.

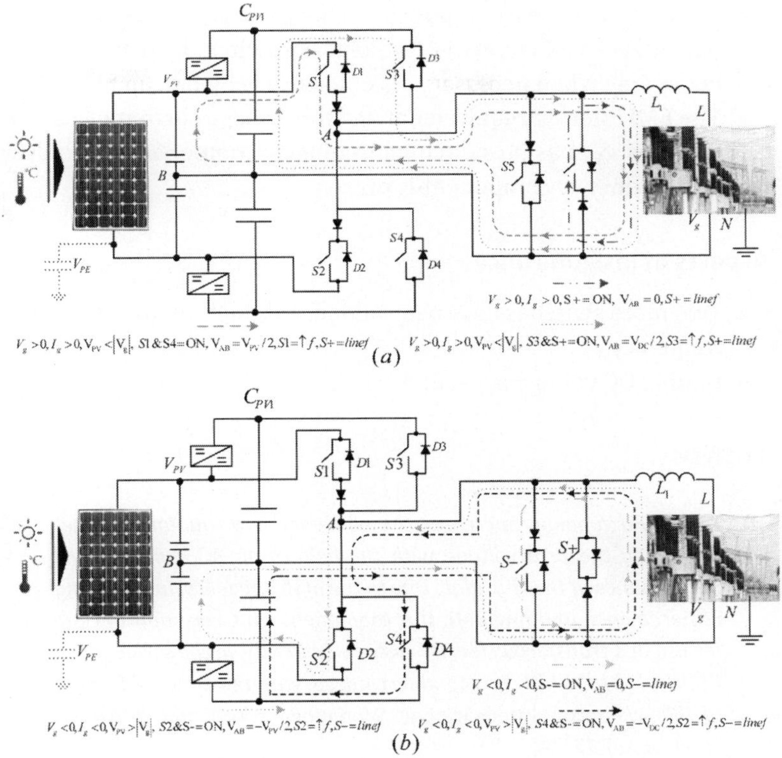

Figure 4.12 Switching states of REFU (REFUSol) in the case of (a) a positive cycle and (b) a negative cycle.

- When there is no need for boost $V_{PV} > |V_g|$, then S1 (S4) is switched at a high frequency.
- When boost is enabled $V_{PV} < |V_g|$, then S3 (S4) is switched at a high frequency.
- Depending on voltage polarity, S+ (S−) is switched at the grid frequency.

Aspects of significance

- The core losses are lower due to UP voltage variation, that is $(0 \rightarrow +V_{PV} \rightarrow 0 \rightarrow -V_{PV} \rightarrow 0)$.

- Because of no reactive power exchange between C_{PV} and $L_{1(2)}$ during zero voltage, lower frequency of switching in one leg, and boost only when necessary, its efficiency is higher, up to 98%.
- The EMI filtering requirement and the leakage current peaks are lower as V_{PE} has no switching frequency component and only a grid frequency component is present.

Aspects of insignificance

- Two more switches have been added, although switched at a low frequency.
- Double DC voltage is needed.

Analysis

> *The REFU topology increases the efficiency with minimum losses by adding the zero voltage with the help of an AC bypass to the performance of the H-bridge. Due to the high efficiency, low filtering requirements, and low EMI, this topology is thus very appropriate for use in a transformerless PV system. Currently, in the series called REFUsol (11/15 kW) it is commercialized, with maximum efficiency of 98% (*Photon International, *September 2008) and European efficiency of 97.5%.*

4.3.5 FB-DCBP (Ingeteam) Inverter

The modification of a classical H-bridge by Ingeteam[29] gives an FB inverter with a DC bypass (FB-DCBP) (patent pending; published by Gonzalez et al.).[30] This topology usually consists of a classical H-bridge with the addition of two extra switches in the DC link, and for clamping the output to the ground, two extra diodes are connected at the middle point of the DC bus, as presented in Fig. 4.13. In conflict to H5 or HERIC, where zero voltage is fluctuating, the clamping diodes ensure that the zero voltage is grounded and the panels are separated from the grid by DC switches during zero-voltage states. During zero voltage, prevention of the reactive power exchange between $C_{PV1(2)}$ and $L_{1(2)}$ leads to high efficiency and low leakage current; essentially "jump-free" V_{PE} solutions are ensured by both. Figure 4.14 presents the generation of AC currents for the positive

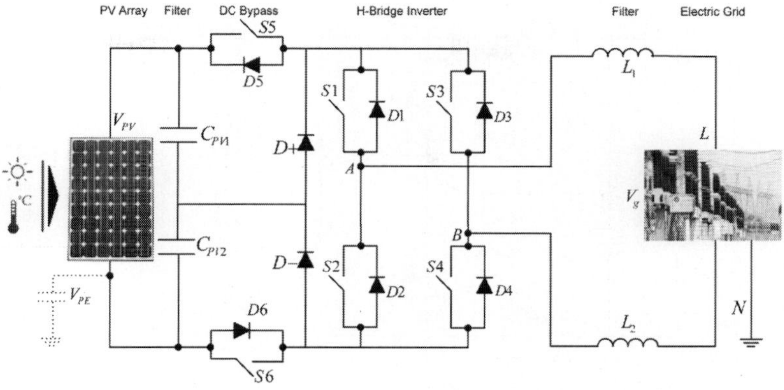

Figure 4.13 Schematic diagram of an FB-DCBP (Ingeteam) inverter.

and negative switching states. The main functions of the FB-DCBP topology are as follows:

- S1 (S2) and S4 (S3) are switched at the grid frequency, and S_5 and S_6 are switched at a high frequency.
- By turning off the DC bypass switches S5 and S6, a zero-output voltage is achieved. The current splits into two paths, when S2 and S3 are turned on and S5 and S6 are turned off, that is, (a) S4 and the freewheeling diode of S2 (D2) and (b) S1 and the freewheeling diode of S3 (D3). Thus, no switching losses appear as S2 and S3 are turned on with no current. The path of the current during the zero-voltage state for a negative grid current will be S2-D4 or S3-D1, while that for the positive grid current will be S1-D3 or S4-D2. To the half of the DC-link voltage, for clamping the bypass switches D+ and D– are used.[9]

Aspects of significance

- The core losses are lower due to UP voltage variation, that is, $(0 \rightarrow +V_{\mathrm{PE}} \rightarrow 0 \rightarrow -V_{\mathrm{PE}} \rightarrow 0)$.
- The DC bypass switches have half the rating of the DC voltage.
- Because of a low voltage rating of S5 and S6, a lower switching frequency in the FB, and no reactive power exchange between C_{PV} and $L_{1(2)}$ during zero voltage, its efficiency is higher.

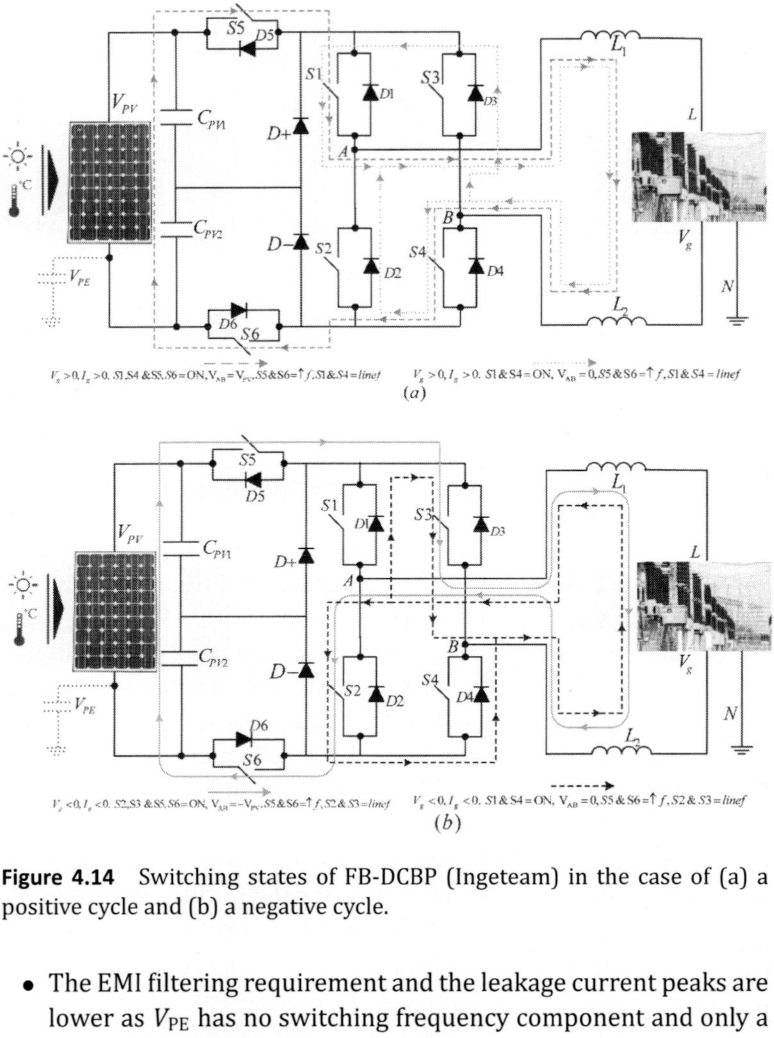

Figure 4.14 Switching states of FB-DCBP (Ingeteam) in the case of (a) a positive cycle and (b) a negative cycle.

- The EMI filtering requirement and the leakage current peaks are lower as V_{PE} has no switching frequency component and only a grid frequency component is present.

Aspects of insignificance

- Two extra diodes and two more switches have been added.
- Without affecting the overall high efficiency, the conduction losses are higher because four switches are conducting during the active vector.

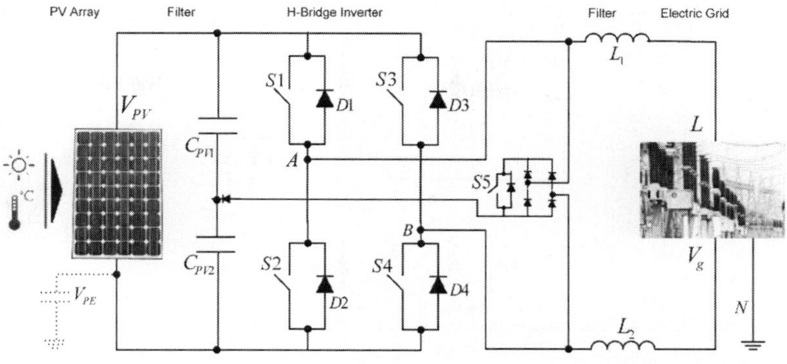

Figure 4.15 Schematic diagram of an FB-ZVR inverter.

Analysis

*FB-DCBP (Ingeteam), due to high efficiency, low filtering requirements, and low EMI in this topology, is thus very appropriate for use in a transformerless PV system. Currently, in the series called Sun TL series (2.5/3.6/6 kW) it is commercialized by Ingeteam in Ingecon, with maximum efficiency of 96.5% (*Photon International, *August 2007) and European efficiency of 95.1%.*

4.3.6 Full-Bridge Zero-Voltage Rectifier Inverter

The other modification of the classical H-bridge is known as the FB zero-voltage rectifier (FB-ZVR),[31] as presented in Fig. 4.15. The derivation of this topology is based on the HERIC topology; using a diode clamp to the DC midpoint and one switch (S5) and a diode bridge, the bidirectional grid short-circuiting switch is implemented. By turning S5 on and turning the FB off a zero voltage is achieved.

Figure 4.16 presents the generation of AC currents for the positive and negative switching states. The main function of the FB-ZVR topology are as follows:

- Like in BP modulation, the switches are diagonally switched in the FB. By turning S5 on and by turning all switches off a bridge in the zero state is introduced.

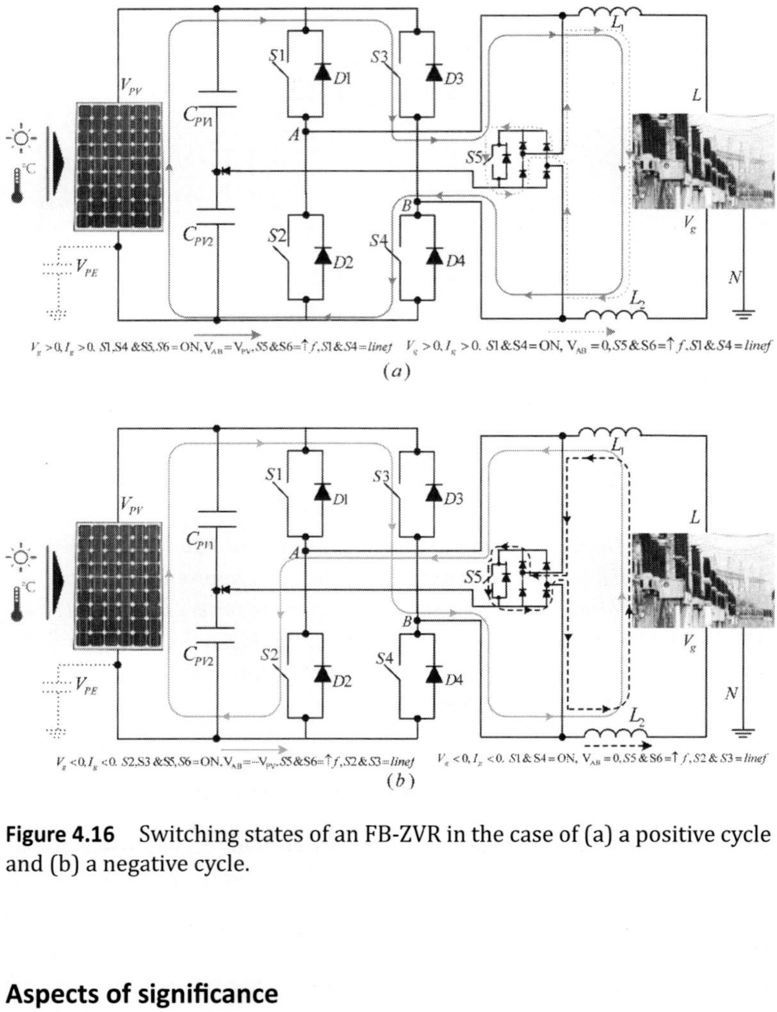

Figure 4.16 Switching states of an FB-ZVR in the case of (a) a positive cycle and (b) a negative cycle.

Aspects of significance

- The core losses are lower due to UP voltage variation, that is, $(0 \rightarrow +V_{PV} \rightarrow 0 \rightarrow -V_{PV} \rightarrow 0)$.
- Because of a lower switching frequency in one leg and no reactive power exchange between C_{PV} and $L_{1(2)}$ during zero voltage, its efficiency is higher, up to 96%.
- The EMI filtering requirement and the leakage current peaks are lower as V_{PE} has no switching frequency component and only a grid frequency component is present.

Aspects of insignificance

- Four extra diodes and one more switch have been added.
- Losses across the filter increase as the BP output is obtained during dead-time clamping.

Analysis

In terms of low leakage and high efficiency the advantages of the HERIC are inherited by the FB-ZVR. The efficiency is lower than that of HERIC because of the high switching frequency of S5; however, it has the advantage that it can work at any power factor.

4.4 FB-Derived Inverter Topologies: An Overview

Actually, the two-level FB (or H-bridge) inverter is converted into a three-level one by using FB-DCBP, REFU, H5, and HERIC topologies. The input voltage stress on both the output inductor and the switches is reduced to half, thus increasing the efficiency. By using an additional DC bypass (FB-DCBP) or AC bypass (REFU or HERIC) or higher switches of the bridge (H5), the zero-voltage state is accomplished by shorting the grid. FB-DCBP and REFU clamp the neutral to the midpoint of the DC link, while HERIC and H5 isolate the grid from the PV panels during zero voltage. Together HERIC and REFU use the AC bypass; however, HERIC uses two switches in series (back to back) and REFU uses two switches in antiparallel. Therefore, for the REFU topology the conduction losses in the AC bypass are lower. The efficiency of H5 and REFU is slightly higher in comparison to FB-DCBP and HERIC as they have only one switch that switches with a high frequency while the others have two.

The implementation of the FB-ZVR is different. Although it's derived from HERIC, it uses one switch and diode bridge as a bidirectional switch. This topology can also work with a nonunitary power factor and have constant V_{PE} but moderate high efficiency (higher than FB-BP but lower than HERIC)

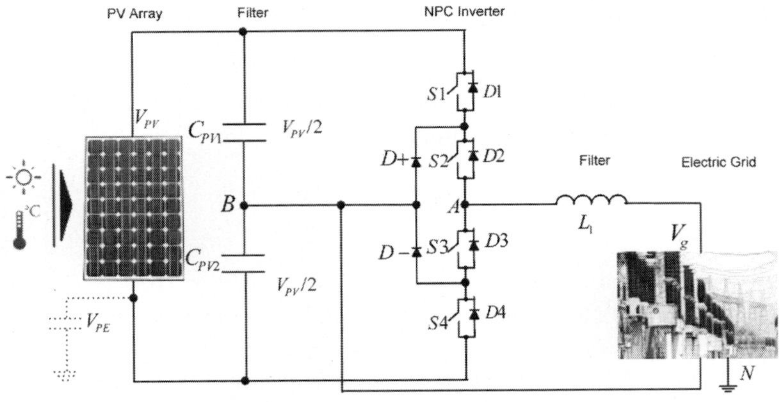

Figure 4.17 Schematic diagram of an FB-ZVR inverter.

4.5 NPC-Based Inverter Structures

NPC was first introduced by Takahashi et al. in 1981.[33] In comparison to a two-level FB inverter, NPC has lower switching stress as well as dV/dt. Because of the versatile nature of NPC topology, it is effective in both three-phase and single-phase (FB or H-bridge) systems.

4.5.1 NPC H-Bridge Inverter

An NPC HF inverter is based on the concept in which zero voltage is acquired by using D+ or D– depending on the sign of the current the output is clamped to at the middle point (ground) of the DC bus, as presented in Fig. 4.17.

Figure 4.18 presents the generation of AC currents for the positive and negative switching states.

The main functions of an NPC HF topology are:

- When there is no need for a boost, $V_{PV} > |V_g|$, the possible zero-voltage states S3, D– = ON and S2, D+ = ON are switched at a high frequency. In out-of-unitary-power-factors operation in opposition for $V_g > 0$ switches S1 and S3 and for $V_g > 0$, $I_g > 0$ switches S2 and S4 are operated.

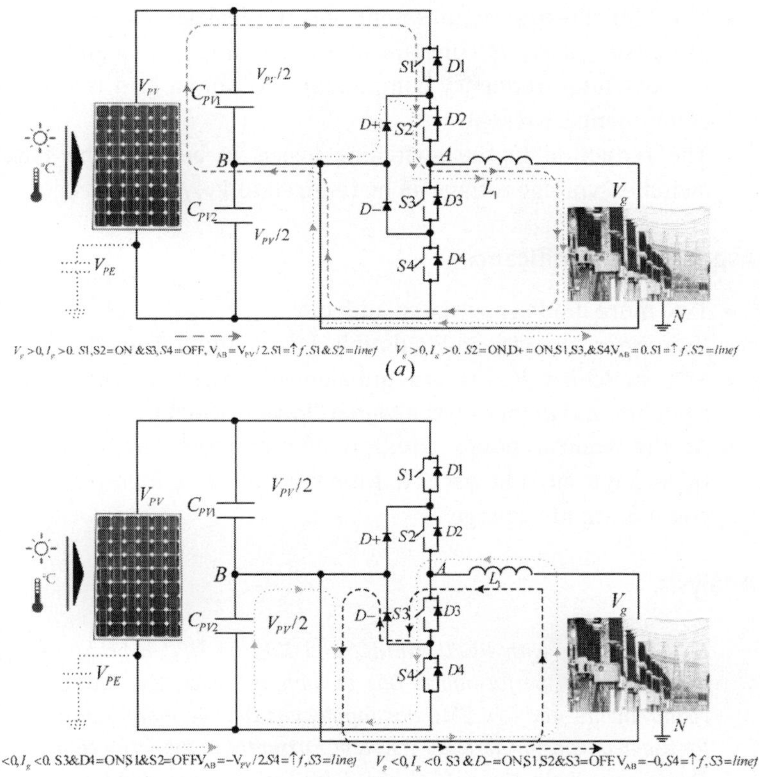

Figure 4.18 Switching states of an NPC H-bridge in the case of (a) a positive cycle and (b) a negative cycle.

- S2 (S3) is switched at the grid frequency, while S1 (S4) is switched at a high frequency.

Aspects of significance

- The core losses are lower due to UP voltage variation, that is, $(0 \rightarrow +V_{PV} \rightarrow 0 \rightarrow -V_{PV} \rightarrow 0)$.
- Because of no reactive power exchange between C_{PV} and $L_{1(2)}$ during zero voltage and lower frequency of switching in one leg, its efficiency is higher, up to 98%.

- The EMI filtering requirement and the leakage current peaks are lower as V_{PE} is constant and is equal to $V_{PV}/4$ and it has no switching frequency component and only a grid frequency component is present.
- The reduction in the switching losses is because the outer switches' voltage rating can be reduced to $V_{PV}/4$.

Aspects of insignificance

- Two more diodes have been added.
- In comparison with an FB, double voltage input is needed.
- The switching losses are unbalanced, lower on the middle switches and higher on the higher/lower switches.
- At the neutral point, addition of any inductance will lead to leakage current as EMI filters generate a high-frequency common-mode voltage.

Analysis

*In comparison with REFU, HERIC, and H5, the NPC H-bridge is very similar in performance. Due to high efficiency, low filtering requirements, and low EMI, this topology is thus very appropriate for use in a transformerless PV system. Currently, in the series called TripleLynx (three-phase 10/12.5/15 kW) it is commercialized by Danfoss Solar Inverter, with maximum efficiency of 98% (*Photon *Magazine, July 2010) and European efficiency of 97%.*

4.5.2 Conergy NPC Inverter

The variation of classical NPC yields to the HF topology whose output is clamped to the neutral using two back-to-back-IGBT-based bidirectional switches. This topology is presented in Fig. 4.19 and patented by Conergy.[33] An alternative of the same concept topology is depicted in the international patent application by Knaup,[34] where instead of an H-bridge an FB is used and in place of a series connection a parallel connection of unidirectional clamping switches is carried out.

An NPC HF inverter is based on the main concept in which zero voltage is acquired by using S+ or S– depending on the sign of the

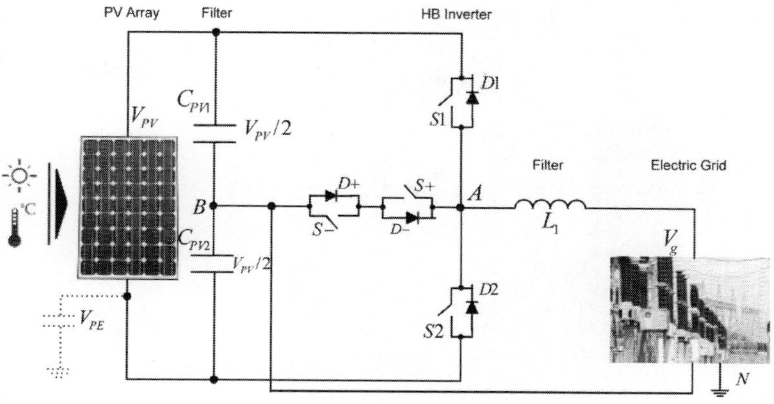

Figure 4.19 Schematic diagram of a Conergy NPC inverter.

current the output is clamped to at the middle point (ground) of the DC bus, as presented in Fig. 4.20. The main functions of the Conergy NPC topology are:

- The possible zero-voltage states are S+, D+ = ON and S–, D– = ON.
- S+ (S–) and S1 (S2) are switched at a high frequency.

Aspects of significance

- The core losses are lower due to UP voltage variation, that is, $(0 \rightarrow +V_{\text{PV}} \rightarrow 0 \rightarrow -V_{\text{PV}} \rightarrow 0)$.
- Because of no reactive power exchange between C_{PV} and $L_{1(2)}$ during zero voltage and because losses are reduced as the on switch is conducting during the active state, its efficiency is higher, up to 98%.
- The EMI filtering requirement and the leakage current peaks are lower as V_{PE} is constant and is equal to $V_{\text{PV}}/4$ and it has no switching frequency component and only a grid frequency component is present.
- In comparison with classical NPC the switching losses are balanced.

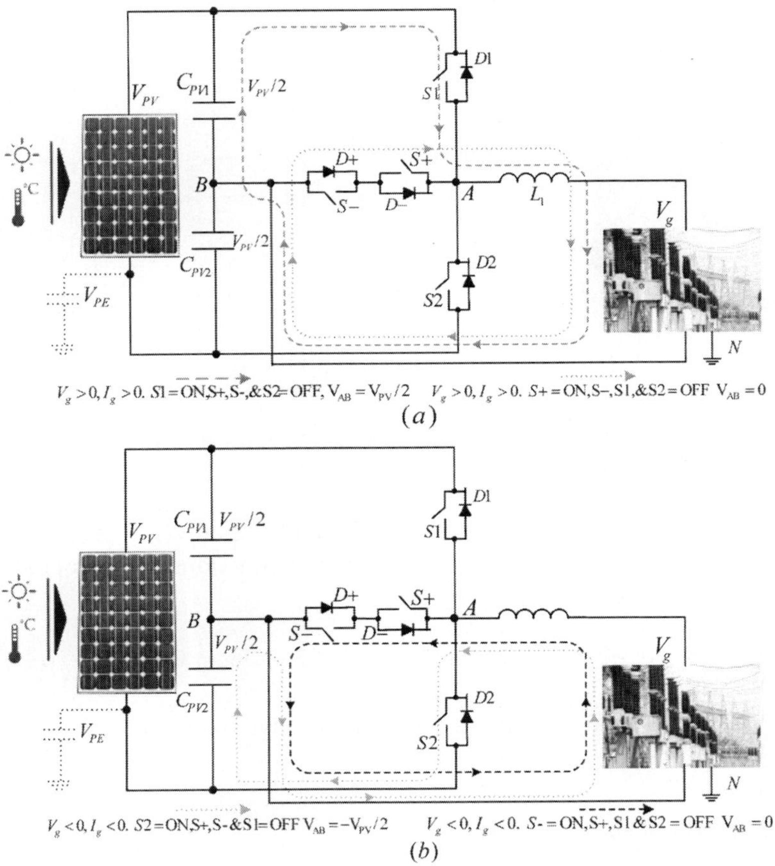

Figure 4.20 Switching states of Conergy NPC in the case of (a) a positive cycle and (b) a negative cycle.

Aspects of insignificance

- In comparison with an FB, double voltage input is required.
- Compared to the outer switches of NPC the voltage rating of S1 and S2 is double.
- At the neutral point, addition of any inductance will lead to leakage current as EMI filters generate a high-frequency common-mode voltage.

Analysis

*Due to slightly higher efficiency compared with classical NPC, low filtering requirements, and low EMI, this topology is very appropriate for use in a transformerless PV system. Currently, in the series called IPG series (2–5 kW) it is commercialized by Conergy in the string inverter series, with maximum efficiency of 96.1% (*Photon International, *July 2007) and maximum European efficiency of 95.1%.*

4.6 NPC-Derived Inverter Topologies: An Overview

All the NPC-derived inverter topologies are three-level topologies. The advantageous features of these topologies are (a) practically no leakage due to a grounded DC link midpoint, (b) high efficiency due to clamping of PV panels during the zero-voltage state, and (c) UP voltage across the filter. In comparison with FB-derived topologies because of higher complexity, with ratings over 10 kW (minicentral) three-phase inverters these topologies are typically used. Besides, for high power, in the range of hundreds of kW (central inverter), where multilevel inverters are important, these topologies are also very attractive.

4.7 Typical PV Inverter Structures

Transformerless PV inverter–based inverter structures are elaborated in this chapter. Conversely, the final structure will be different as most of them require boosting. Since the power of a single PV panel is low and strongly dependent on ambient temperature and solar irradiance (ambient conditions), either a boost or a buck-boost converter is required to achieve an acceptable DC-link voltage.[35–41]

Figure 4.21 presents an H-bridge boosting PV inverter with a high-frequency transformer, one with a low frequency transformer, and one without a transformer. Additionally, high-frequency versions (HERIC or H5) can easily replace the FB inverter.

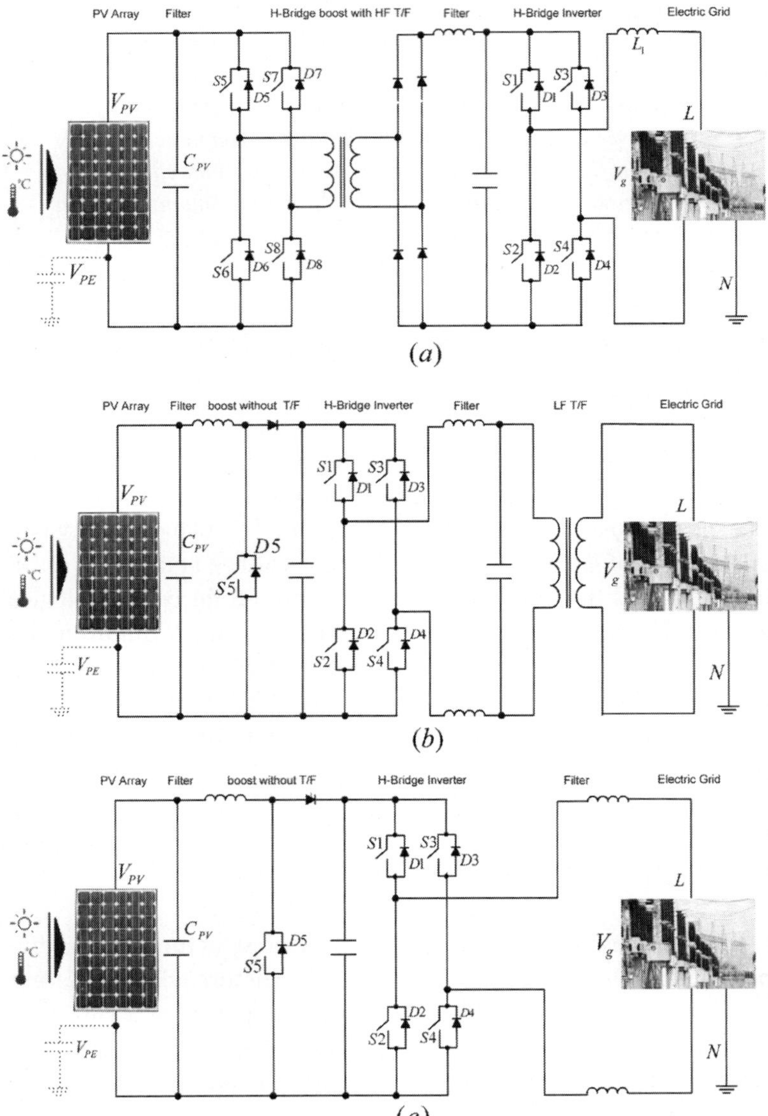

Figure 4.21 H-bridge-based PV inverter with (a) a high-frequency transformer, (b) a low-frequency transformer, and (c) without a transformer.

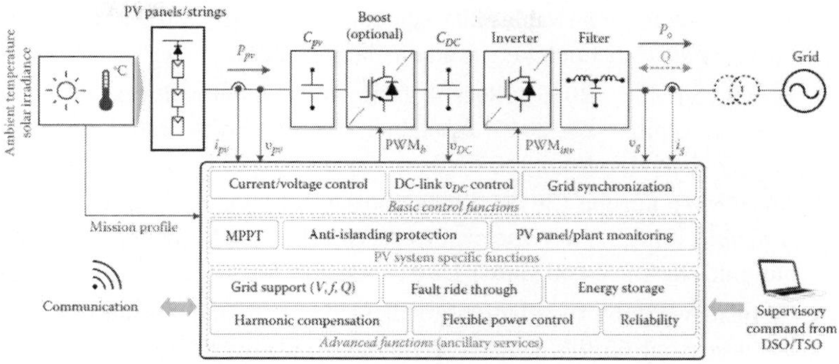

Figure 4.22 Generic control structure of a single-phase grid-connected system.

4.8 Generic Control Structure for a Single-Phase Grid-Connected System

The main purposes of the control for SPG-C systems are as follows: (a) to maximize the power from PV panels, PV-side control is incorporated, and (b) for the purpose of fulfilling the demands of the power grid, grid-side control is performed. To fulfill the requirements/demands the generic control structure consists of two cascaded loops.[42,43] The inner control loop is accountable for shaping the current, while the outer voltage/power control loop produces the current command. In this way the quality of power is sustained along with various other functionalities, as depicted in Fig. 4.22.

References

1. O. Ellabban, H. Abu-Rub, F. Blaabjerg (2014). Renewable energy resources: current status, future prospects and their enabling technology, *Renewable Sustainable Energy Rev.*, **39**, pp. 748–764.
2. L. Hassaine, E. OLias, J. Quintero, V. Salas (2014). Overview of power inverter topologies and control structures for grid connected photovoltaic systems, *Renewable Sustainable Energy Rev.*, **30**, pp. 796–807.

3. REN21 (2015). Renewables 2015 global status report, 2015. Available: http://www.ren21.net. Retrieved on March 1, 2016.
4. Fraunhofer ISE (2015). Recent facts about photovoltaics in Germany, December 25, 2015. Available: http://www.ise.fraunhofer.de/. Retrieved on March 1, 2016.
5. SolarPower Europe (2015). Global market outlook 2015–2019, 2015. Available: http://www.solarpowereurope.tv/insights/global-market-outlook/. Retrieved on March 1, 2016.
6. PV-Insider (2016). US solar jobs to hike 15% in 2016; Global new capacity forecast at 65 GW, January 19, 2016. Available: http://analysis.pv-insider. com/us-solar-jobs-hike-15-2016-global-new-capacity-forecast-65-gw. Retrieved on March 1, 2016.
7. F. Blaabjerg, K. Ma, Y. Yang (2014). Power electronics: the key technology for renewable energy systems, in *Proceedings of EVER*, pp. 1–11.
8. J. D. van Wyk, F. C. Lee (2013). On a future for power electronics, *IEEE J. Emerging Sel. Top. Power Electron.*, **1**(2), pp. 59–72.
9. S. B. Kjaer, J. K. Pedersen, F. Blaabjerg (2005). A review of single-phase grid-connected inverters for photovoltaic modules, *IEEE Trans. Ind. Appl.*, **41**(5), pp. 1292–1306.
10. H. J. Bergveld, D. Buthker, C. Castello, T. Doorn, A. de Jong, R. Van Otten, K. de Waal (2013). Module-level DC/DC conversion for photovoltaic systems: the delta-conversion concept, *IEEE Trans. Power Electron.*, **28**(4), pp. 2005–2013.
11. D. Meneses, F. Blaabjerg, O. Garcia, J. A. Cobos (2013). Review and comparison of step-up transformerless topologies for photovoltaic AC-module application, *IEEE Trans. Power Electron.*, **28**(6), pp. 2649–2663.
12. B. Liu, S. Duan, T. Cai (2011). Photovoltaic DC-building-module-based BIPV system: concept and design considerations, *IEEE Trans. Power Electron.*, **26**(5), pp. 1418–1429.
13. D. Sera, L. Mathe, F. Blaabjerg (2014). Distributed control of PV strings with module integrated converters in presence of a central MPPT, in *Proceedings of IEEE ECCE*, pp. 1–8.
14. M. A. Eltawil, Z. Zhao (2010). Grid-connected photovoltaic power systems: technical and potential problems—a review, *Renewable Sustainable Energy Rev.*, **14**(1), pp. 112–129.
15. F. Spertino, G. Graditi (2014). Power conditioning units in grid-connected photovoltaic systems: a comparison with different technologies and wide range of power ratings, *Sol. Energy*, **108**, pp. 219–229.

16. Solarpraxis and Sunbeam Communications (2013). Inverter, storage and PV system technology 2013, July 2013. Available: http://www.pv-system-tech.com/. Retrieved on March 1, 2016.
17. E. Romero-Cadaval, G. Spagnuolo, I. Garcia Franquelo, C. A. Ramos-Paja, T. Suntio, W. M. Xiao (2013). Grid-connected photovoltaic generation plants: components and operation, *IEEE Ind. Electron. Mag.*, **7**(3), pp. 6–20.
18. M. Meinhardt, G. Cramer (2016). Multi-string-converter: the next step in evolution of string-converter technology, in *Proceedings of EPE*, pp. p.1–p.9, Graz, Austria, 2001. SMA, SUNNY CENTRAL—high tech solution for solar power stations, products category. Available: http://www.sma-america.com/. Retrieved on March 1, 2016.
19. M. Morjaria, D. Anichkov, V. Chadliev, S. Soni (2014). A grid-friendly plant: the role of utility-scale photovoltaic plants in grid stability and reliability, *IEEE Power Energy Mag.*, **12**(3), pp. 87–95.
20. Y. Xue, K. C. Divya, G. Griepentrog, M. Liviu, S. Suresh, M. Manjrekar (2011). Towards next generation photovoltaic inverters, in *Proceedings of IEEE ECCE*, pp. 2467–2474.
21. Q. Li, P. Wolfs (2008). A review of the single phase photovoltaic module integrated converter topologies with three different DC link configurations, *IEEE Trans. Power Electron.*, **23**(3), pp. 1320–1333.
22. Y. Yang, P. Enjeti, F. Blaabjerg, H. Wang (2015). Wide-scale adoption of photovoltaic energy: grid code modifications are explored in the distribution grid, *IEEE Ind. Appl. Mag.*, **21**(5), pp. 21–31.
23. R. Teodorescu, M. Liserre, P. Rodriguez (2011). *Grid Converters for Photovoltaic and Wind Power Systems*, Wiley, Hoboken, NJ.
24. F. Blaabjerg, D. M. Ionel (2017). *Renewable Energy Devices and Systems with Simulations in MATLAB and ANSYS*, Taylor & Francis.
25. W. Mcmurray (1965). Inverter circuits, US Patent 3207974.
26. R.-S. Lai, K. D. T. Ngo (1995). A PWM method for reduction of switching loss in a full-bridge inverter, *IEEE Trans. Power Electron.*, **10**(3), pp. 326–332.
27. M. Victor, et al. (2005). US Patent Application, Publication Number US 2005/0286281 A1.
28. H. Schmid, et al. (2006). US Patent 7046534.
29. J. Hantschel (2007). German Patent Application, Publication Number DE102006010694 A11.
30. S. R. Gonzalez, et al. (2007). International Patent Application, Publication Number WO2008015298.

31. R. Gonzalez, J. Lopez, P. Sanchis, L. Marroyo (2007). Transformerless inverter for single-phase photovoltaic systems, *IEEE Trans. Power Electron.*, **22**(2), pp. 693–697.
32. T. Kerekes, R. Teodorescu, P. Rodriguez, G. Vazquez, E. Aldabas (2010). A new high-efficiency single-phase transformerless PV inverter topology, *IEEE Trans. Ind. Electron.*, **58**, pp. 184–191.
33. A. Nabae, H. Magi, I. Takahashi (1981). A new neutral-point-clamped PWM inverter, *IEEE Trans. Ind. Appl.*, **IA-17**(5), pp. 518–523.
34. P. Knaup (2007). International Patent Application, Publication Number WO 2007/048420 A1.
35. M. Calais, V. G. Agelidis, M. Meinhardt (1999). Multilevel converters for single-phase grid connected photovoltaic systems: an overview, *Sol. Energy*, **66**(5), pp. 325–335.
36. B. S. Prasad, S. Jain, V. Agarwal (2008). Universal single-stage grid-connected inverter, *IEEE Trans. Energy Convers.*, **23**(1), pp. 128–137.
37. L. G. Junior, M. A. G. de Brito, L. P. Sampaio, C. A. Canesin (2011) Single stage converters for low power stand-alone and grid-connected PV systems, in *Proceedings of IEEE ISIE*, pp. 1112–1117.
38. R. O. Caceres, I. Barbi (1999). A boost DC–AC converter: analysis, design, and experimentation, *IEEE Trans. Power Electron.*, **14**(1), pp. 134–141.
39. S. Funabiki, T. Tanaka, T. Nishi (2002). A new buck-boost-operation-based sinusoidal inverter circuit, in *Proceedings of IEEE PESC*, pp. 1624–1629.
40. N. Vazquez, J. Almazan, J. Alvarez, C. Aguilar, J. Arau (1999). Analysis and experimental study of the buck, boost and buck-boost inverters, in *Proceedings of IEEE PESC*, pp. 801–806.
41. C. Wang (2004). A novel single-stage full-bridge buck-boost inverter, *IEEE Trans. Power Electron.*, **19**(1), pp. 150–159.
42. S. Jain, V. Agarwal (2007). A single-stage grid connected inverter topology for solar PV systems with maximum power point tracking, *IEEE Trans. Power Electron.*, **22**(5), pp. 1928–1940.
43. F. Blaabjerg, R. Teodorescu, M. Liserre, A. V. Timbus (2006). Overview of control and grid synchronization for distributed power generation systems, *IEEE Trans. Ind. Electron.*, **53**(5), pp. 1398–1409.

Chapter 5

Facile One-Step Synthesis of a Composite CuO/Co_3O_4 Electrode Material on Ni Foam for Flexible Supercapacitor Applications

In this chapter, for the first time, the composite CuO/Co_3O_4 electrode material is explored for supercapacitor applications. We have synthesized a Co_3O_4-decorated and CuO-decorated Co_3O_4 (CuO/Co_3O_4) composite material by a facile hydrothermal method. Compared to a Co_3O_4 electrode, the CuO/Co_3O_4 electrode delivered a high specific capacitance of 806.25 F g^{-1} at 2 A g^{-1} and superior cycling stability, with the retention of 99.75% after 2000 cycles. Excitingly, the CuO/Co_3O_4 electrode shows excellent electrochemical performance under various bending angles, representing superior flexibility.

5.1 Overall Survey

Nowadays, supercapacitors are considered as more promising energy storage systems compared to batteries and traditional

Solar Power and Energy Storage Systems
Hee-Je Kim

ISBN 978-981-4800-13-6 (Hardcover), 978-0-429-45877-4 (eBook)
www.panstanford.com

capacitors, due to their fast charge-discharge rates, high energy density and power density, long cycling stability, and low cost.[1] The performance of an electrochemical supercapacitor is mainly dependent on the electrode material. In general, supercapacitors are classified into electrochemical double-layer capacitors (EDLCs) and pseudocapacitors on the basis of application of active materials and electrochemical charge-discharge behavior.[2] In fact, pseudo-capacitors show higher peformance than EDLCs owing to their fast diffusion redox reactions. Therefore, a lot of research effort is directed toward the development of low-cost and high-performance electrode materials for further improving the performance of pseudocapacitors.[3]

In recent years, transition-metal oxides (Co_3O_4, $MnCo_2O_4$, MnO_2, Fe_2O_3, CuO, $NiCo_2O_4$, Co_3O_4/CuO, $Ni_{0.8}Co_{0.1}Mn_{0.1}O_{1.1}$, etc.) have been extensively proposed for SCs and lithium-ion-battery applications, due to their excellent electroactive abilities.[4–13] Among them, CuO and Co_3O_4 are most widely used electrode materials due to their low cost, environmental friendliness, and abundance, as well as their high theoretical specific capacitance.[14–17] In this work, for the first time, composite CuO/Co_3O_4 nanostructures were synthesized on Ni foam via a facile one-step hydrothermal route. Interestingly, the CuO/Co_3O_4 composite electrode delivered a high specific capacitance of 806.25 F g^{-1} at 2 A g^{-1} and promising long-term cyclic stability, with only 0.25% loss of specific capacity over 1000 cycles.

5.2 Materials and Methods

Scanning electron microscope (SEM) morphologies of Co_3O_4 and CuO/Co_3O_4 are shown in Fig. 5.1. The Co_3O_4 electrode shows agglomerated nanospheres (Fig. 5.1a), and a high-magnification Co_3O_4 SEM image reveals a sphere morphology with a diameter of 1.8 μm (Fig. 5.1b). The SEM image of CuO/Co_3O_4 shows a sphere morphology with a size of about 0.6–1.4 μm (Fig. 5.1c). As seen from the high-magnification SEM image (Fig. 5.1d), CuO nanoflakes were grown on the Co_3O_4 core.

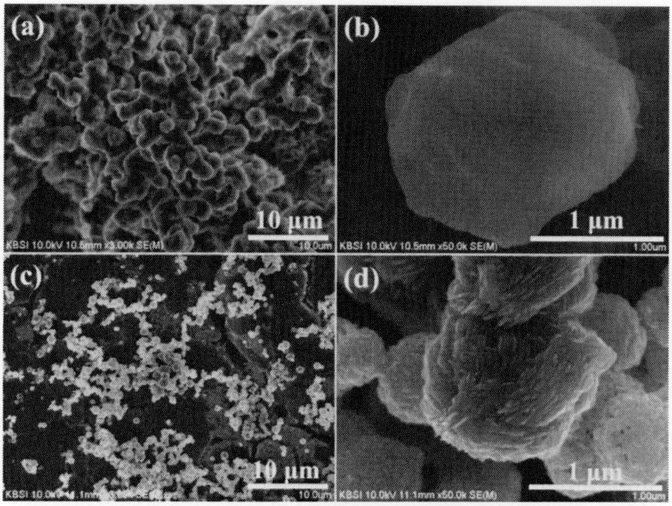

Figure 5.1 SEM images of (a, b) Co_3O_4 and (c, d) CuO/Co_3O_4-coated Ni foam electrodes.

Moreover, SEM-energy-dispersive X-ray (EDX) mapping analysis was conducted to identify the uniformity and interconnection of CuO and Co_3O_4 on the Ni foam surface (Fig. 5.2), where Cu, Co, and O elements are uniformly distributed in the sample. However, the map of Cu shows that its distribution is sporadic, implying that only a small amount of CuO is exposed on the surface of Co_3O_4.

The growth mechanism of CuO/Co_3O_4 is discussed here: at the starting of the hydrothermal reactions, in the homogeneous solution, Co^{2+} and Cu^{2+} ions can react with F^- ions and form CoF^+ and CuF^+ complexes. When the temperature reaches 100°C, urea releases the OH^- and carbonate ions during hydrolysis. Then, CuF^+ and OH^- ions form a Cu(OH)F precipitate on the surface of the Ni foam. When the CuF^+ ions are consumed to some degree, the CoF^+ ions begin to precipitate into nanoflakes on the Cu(OH)F core. The formation mechanism can be expressed as the following reactions:[15]

$$CO(NH_2)_2 + H_2O \rightarrow 2NH_3 + CO_2 \tag{5.1}$$

$$H_2O + NH_3 \rightarrow OH^- + NH^{4+} \tag{5.2}$$

$$CuF^+ + OH^- \rightarrow Cu(OH)F \tag{5.3}$$

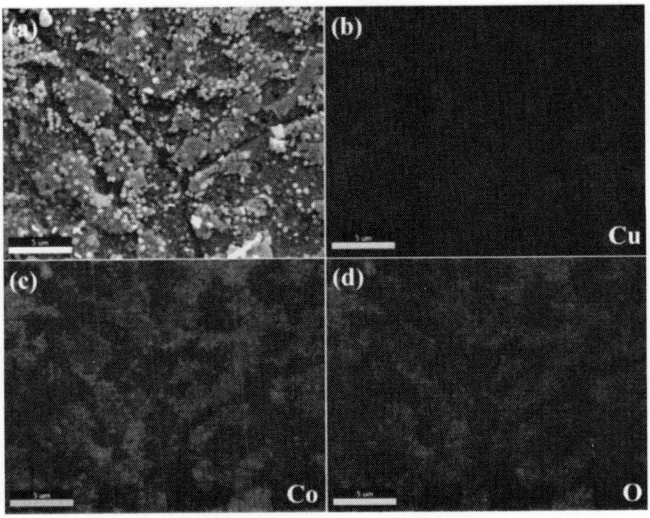

Figure 5.2 SEM-EDX mapping results for Cu, Co, and O elements in a CuO/Co$_3$O$_4$ electrode.

$$Co_2 + H_2O \rightarrow CO_3^{2-} + 2H^+ \tag{5.4}$$

$$CoF^+ + xOH^- + 0.5(2-x)CO_3^{2-} + nH_2O \rightarrow Co(OH)_x(CO_3)_{0.5(2-x)} \cdot nH_2O + F^- \tag{5.5}$$

$$Cu(OH)F \rightarrow CuO + HF \uparrow \tag{5.6}$$

$$Co(OH)_x(CO_3)_{0.5(2-x)} \cdot nH_2O + O_2 \rightarrow Co_3O_4 + Co_2 \uparrow + H_2O \uparrow \tag{5.7}$$

Furthermore, the crystal structure and phase of the electrodes were studied by X-ray diffraction (XRD) analysis. Figure 5.3 depicts the XRD pattern of the composite CuO/Co$_3$O$_4$ electrode on the surface of Ni foam. Besides the Ni foam substrate peaks, the peaks can be indexed to the (11-1), (111), and (31-1) planes of the CuO phase (JCPDS 48-1548)[18] and to the (311), (511), and (440) planes of the cubic Co$_3$O$_4$ (JCPDS No. 42-1467).[15] The SEM-EDX mapping and XRD results confirm that CuO/Co$_3$O$_4$ is successfully deposited on the Ni foam surface.

Cyclic voltammetry (CV), galvanostatic charge-discharge (GCD), and electrochemical impedance spectroscopy (EIS) are appropriate

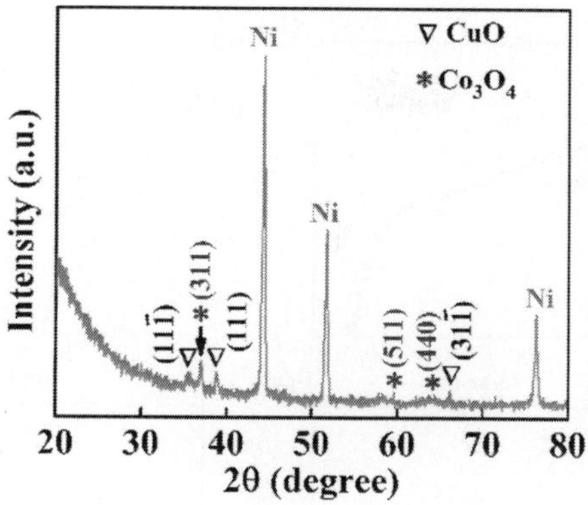

Figure 5.3 XRD pattern of a composite CuO/Co_3O_4 sample on a Ni foam surface.

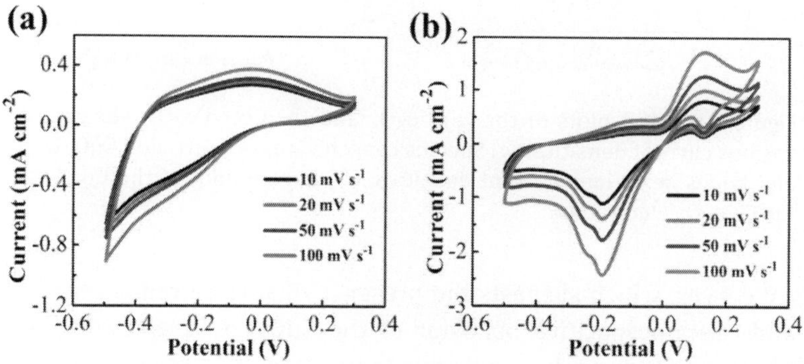

Figure 5.4 CV plots of the (a) Co_3O_4 and (b) CuO/Co_3O_4 electrodes at various scan rates.

tools to demonstrate the electrochemical behavior of electrodes in a three-electrode system with a 3 M KOH electrolyte. Figure 5.4 depicts the CV plots of the Co_3O_4 and CuO/Co_3O_4 samples recorded at various scan rates in the potential range of −0.5 to +0.3 V. The shapes of the CV curves of Co_3O_4 and CuO/Co_3O_4 are clearly different from the ideal rectangular shapes related to the electric double-layer capacitor.[19] This indicates that the as-prepared Co_3O_4 and CuO/Co_3O_4 electrodes are excellent potential capacitive materials with pseudocapacitance. The current density is increasing with an increase in the scan rate from 10 to 100 mV s^{-1} in the

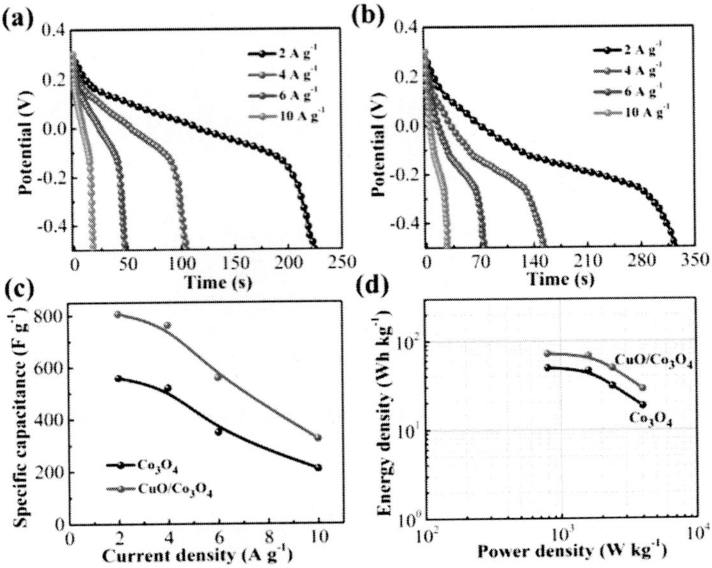

Figure 5.5 GCD plots of the (a) Co_3O_4 and (b) CuO/Co_3O_4 electrodes at various current densities. (c) Specific capacitances of Co_3O_4 and CuO/Co_3O_4 electrodes at various current densities. (d) Ragone plot of the Co_3O_4 and CuO/Co_3O_4 electrodes.

CV curves, which suggests the presence of surface redox reactions and pseudocapacitive behavior of the CuO/Co_3O_4 electrode. The observed redox peaks can be due to the reversible electron transfer process that occurred at the CuO/Co_3O_4 electrode.[20] The CV areas enlarged with an increasing scan rate, denoting good capacitance retention. However, the CuO/Co_3O_4 sample delivered higher current under the curve and a significantly larger area with prominent redox peaks compared to the Co_3O_4 electrodes, indicating a larger capacitance observed for the CuO/Co_3O_4 electrode.

Figure 5.5 depicts the GCD curves of Co_3O_4 and CuO/Co_3O_4 electrodes at various current densities in the potential window of −0.5 to +0.3 V. Both electrodes exhibited similar charge-discharge behavior, but the duration of charging-discharging was different.

A CuO/Co_3O_4 electrode shows longer discharging duration than Co_3O_4, indicating the substantially better performance of CuO/Co_3O_4. Figure 5.5c depicts the specific capacitance as a function

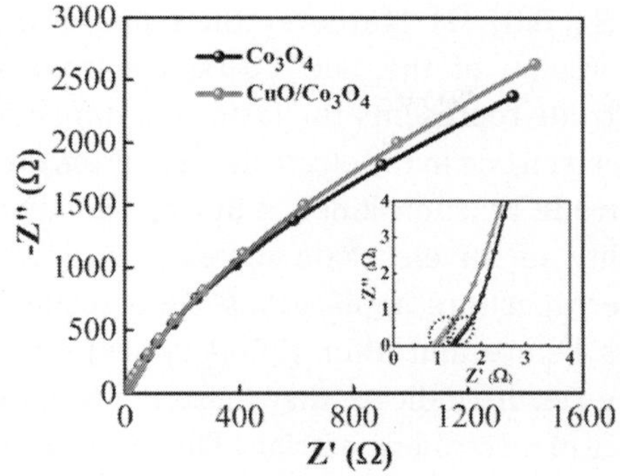

Figure 5.6 EIS plots of Co_3O_4 and CuO/Co_3O_4 electrodes. The inset depicts the enlarged EIS of the electrodes.

of current density, and the specific capacitance decreases with an increase in the current density. The Co_3O_4 sample delivers low specific capacitance values of 559.42, 520.62, 350.25, and 209.62 F g^{-1} at current densities of 2, 4, 6, and 10 A g^{-1}, respectively. Besides, a CuO/Co_3O_4 electrode exhibits higher specific capacitance values of 806.25, 762.5, 559.28, and 325 F g^{-1} at the same current densities. Furthermore, a Ragone plot was used to investigate the performance of electrodes. The CuO/Co_3O_4 electrode delivered the maximal energy density of 71.66 W h kg^{-1} and a power density of 800 W kg^{-1}, while the Co_3O_4 sample exhibited the lower energy density of 49.72 W h kg^{-1} and a power density of 799.89 W kg^{-1} (Fig. 5.5d). The synergistic interaction of CuO and Co_3O_4 facilitates the electrolyte transfer at the electrode interfaces, which indicates the superiority of the composite electrode compared to bare Co_3O_4.

EIS was investigated to demonstrate the electrochemical properties of the electrodes. Figure 5.6 shows Nyquist plots of Co_3O_4 and CuO/Co_3O_4 electrodes. The impedance plot illustrates a small semicircle at a high-frequency region and a straight line at a low-frequency region. The intercept of the plot at the real axis (Z′) denotes the equivalent series resistance (ESR), and the small semicircle stands for the charge-transfer resistance (R_{ct}) at the electrode/electrolyte interface. Due to facile ion transport in an electrolyte, the ESR (0.68 Ω) and R_{ct} (0.56 Ω) values of the CuO/Co_3O_4 electrode are lower than that of the CuO electrode

(ESR, 0.94 Ω; R_{ct}, 0.87 Ω). Moreover, the more vertical line in the low-frequency region of the CuO/Co_3O_4 electrode compared to the Co_3O_4 electrode represents the faster ion diffusion behavior of the CuO/Co_3O_4 electrode in the electrolyte solution, thereby yielding improved electrode reaction kinetics during the charge-discharge process. Stability of an electrode material is a very important factor for supercapacitors applications. As a result, the CV, GCD, and EIS curves before and after (2000 cycles) are illustrated in Fig. 5.7a–c. The results indicate that there is no significant loss in the performance of electrode materials. The EIS results indicate that the small change in ESR (0.68 Ω to 0.70 Ω) is due to dissolution of the CuO/Co_3O_4 material during the repeated charge-discharge process. The results reveal that the CuO/Co_3O_4 electrode exhibits excellent cycling stability. The cycling stability of the Co_3O_4 and CuO/Co_3O_4 electrodes was measured at a current density of 6 A g^{-1} for 2000 cycles, as shown in Fig. 5.7d. The capacity retention ratios of Co_3O_4 and CuO/Co_3O_4 are 94.21% and 99.75%, respectively, certifying the outstanding stability of a CuO/Co_3O_4 electrode. The electrochemical performance of the CuO/Co_3O_4 electrode is further examined under bending conditions. Figure 5.8 depicts that the CV and GCD curves of the CuO/Co_3O_4 electrode exhibit no apparent change at different bending angles, indicating excellent flexibility and mechanical stability of the CuO/Co_3O_4 electrode in practical applications. To improve the performance of a CuO/Co_3O_4 composite electrode for supercapacitor applications, search for more detailed analysis and measuring methods is in progress.

5.3 Experiments

5.3.1 Materials Preparation of Co_3O_4 and CuO/Co_3O_4

Analytical-grade chemicals were received from Sigma-Aldrich and used without any further purification. CuO/Co_3O_4 nanocomposites were prepared on Ni foam substrates by a facile hydrothermal route, followed by a simple sintering process. In a typical synthesis process, Ni foam (1 × 1 cm^2) was etched with a 2 M HCl and rinsed with acetone, ethanol, and deionized (DI) water. 0.04 M of $Co(NO_3)_2 \cdot 6H_2O$, 0.02 M of $Cu(NO_3)_2 \cdot 5H_2O$, 0.2 M of urea, and 0.1 M

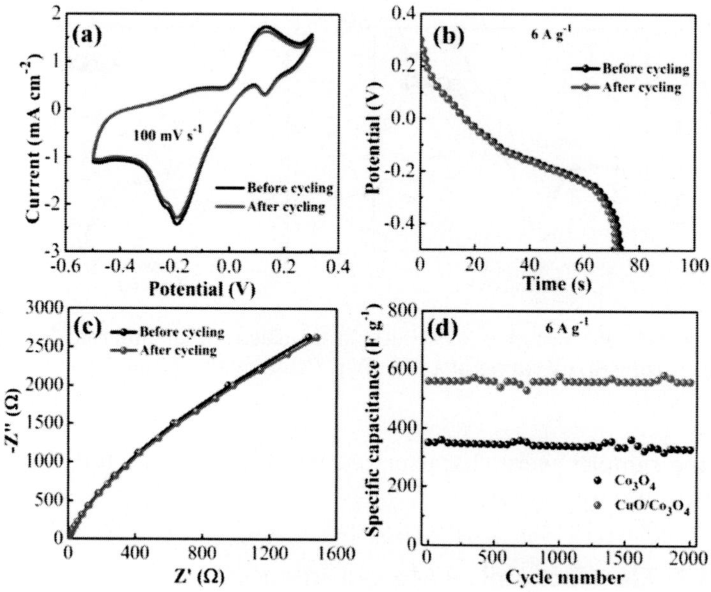

Figure 5.7 Long-term cycling performance of the CuO/Co_3O_4 electrode before and after 2000 cycles: (a) CV plots at a 100 mV s^{-1} scan rate, (b) GCD curves at 6 A g^{-1}, (c) EIS spectra, and (d) cyclic behavior of Co_3O_4 and CuO/Co_3O_4 electrodes at 6 A g^{-1}.

of NH_4F were dissolved in 70 mL of H_2O under constant magnetic stirring. The recepe solution was then transferred into the autoclave liners, and the Ni foam was dipped into the solution, followed by heating to 100°C for 6 h. After cooling to room temperature the active material–loaded Ni foams were taken out and cleaned with DI water. Finally, the electrodes were sintered at 300°C for 2 h in air. The Co_3O_4 nanostructures were fabricated in the same conditions without the addition of $Cu(NO_3)_2{\cdot}5H_2O$. The mass loadings of CuO/Co_3O_4 and Co_3O_4 active material on Ni foam were found to be 1.5 and 1 mg cm^{-2}, respectively.

5.3.2 Characterization

The morphology of the electrodes was analyzed by SEM (S-2400, Hitachi) combined with an EDX spectrometer. The crystal structures

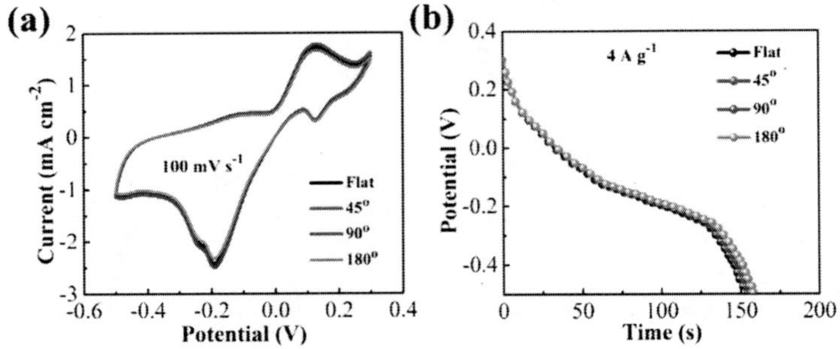

Figure 5.8 Performance of a CuO/Co_3O_4 electrode at different bending angles using (a) CV curves at a 100 mV s^{-1} scan rate and (b) GCD plots taken at 4 A g^{-1}.

of the samples were characterized by XRD (D/Max-2400, Rigaku) with Cu Kα irradiation.

5.3.3 Electrochemical Measurements

CVs, GCD, and EIS characterizations were measured by BioLogic-SP150 workstation at room temperature using a three-electrode system in which CuO/Co_3O_4 and Co_3O_4 were used as working electrodes, a platinum wire counterelectrode, and a standard calomel electrode. The electrolyte consists of a 3 M KOH aqueous solution. EIS analysis was conducted in the frequency range from 100 mHz to 500 kHz. The specific capacitance (C_s), energy density (E), and power density (P) were extracted by the following equations:[18,19]

$$C_s = \frac{I_d \times \Delta t}{m \times \Delta V} \tag{5.8}$$

$$E = \frac{C_s \times (\Delta V^2)}{2} \tag{5.9}$$

$$P = \frac{E}{t} \tag{5.10}$$

Here I_d, t, ΔV, and m denote the constant current density, discharging time, potential window, and mass of the active materials, respectively.

5.4 Conclusions

In summary, we have reported facile and cost-effective hydrothermal synthesis of a composite CuO/Co_3O_4 electrode material for supercapacitor applications. The CuO/Co_3O_4 electrode exhibited a maximum specific capacitance of 806.25 F g^{-1} at 2 A g^{-1} and increased cycling stability of 99.75% retention after 2000 cycles compared to a Co_3O_4 electrode. Besides, the CuO/Co_3O_4 electrode delivered excellent flexibility at different bending angles. The improved performance of the CuO/Co_3O_4 electrode was attributed to the improved surface morphology with a high surface area, which facilitates electron diffusion at the electrode/electrolyte interface. These results suggest that a low-cost, flexible CuO/Co_3O_4 electrode with outstanding electrochemical properties is a potential candidate for supercapacitor applications.

References

1. S. Boukhalfa, K. Evanoff, G. Yushin (2012). *Energy Environ. Sci.*, **5**, pp. 6872–6879.
2. Y. Shao, M. F. El-Kady, L. J. Wang, Q. Zhang, Y. Li, H. Wang, M. F. Mousavi, R. B. Kaner (2015). *Chem. Soc. Rev.*, **44**, pp. 3639–3665.
3. T. Y. Wei, C. H. Chen, H. C. Chien, S. Y. Lu, C. C. Hu (2010). *Adv. Mater.*, **22**, pp. 347–351.
4. Y. Wang, Y. Lei, J. Li, L. Gu, H. Yuan, D. Xiao (2014). *ACS Appl. Mater. Interfaces*, **6**, pp. 6739–6747.
5. J. Gomez, E. E. Kalu (2013). *J. Power Sources*, **230**, pp. 218–224.
6. G. S. Gund, D. P. Dubal, N. R. Chodankar, J. Y. Cho, P. Gomez-Romero, C. Park, C. D. Lokhande (2015). *Sci. Rep.*, **5**, 12454 (13 pp).
7. T. Li, X. Li, Z. Wang, H. Guo, Y. Li (2015). *J. Mater. Chem. A*, **3**, pp. 11970–11975.
8. J. Wang, Q. Zhang, X. Li, D. Xub, Z. Wang, H. Guo, K. Zhang (2014). *Nano Energy*, **6**, pp. 19–26.
9. T. Li, X. Li, Z. Wang, H. Guo (2017). *J. Power Sources*, **342**, pp. 495–503.
10. Q. Zhang, J. Wang, D. Xu, Z. Wang, X. Li, K. Zhang (2014). *J. Mater. Chem. A*, **2**, pp. 3865–3874.

11. J. Wang, Q. Zhang, X. Li, B. Zhang, L. Mai, K. Zhang (2015). *Nano Energy*, **12**, pp. 437–446.
12. Q. Zhang, J. Wang, J. Dong, F. Ding, X. Li, B. Zhang, S. Yang, K. Zhang (2015). *Nano Energy*, **13**, pp. 77–91.
13. Q. Zhang, B. Zhao, J. Wang, C. Qu, H. Sun, K. Zhang, M. Liu (2016). *Nano Energy*, **28**, pp. 475–485.
14. S. E. Moosavifard, M. F. El-Kady, M. S. Rahmanifar, R. B. Kaner, M. F. Mousavi (2015). *ACS Appl. Mater. Interfaces*, **7**, pp. 4851–4860.
15. D. Cai, H. Huang, D. Wang, B. Liu, L. Wang, Y. Liu, Q. Li, T. Wang (2014). *ACS Appl. Mater. Interfaces*, **6**, pp. 15905–15912.
16. J. Wang, Q. Zhang, X. Li, D. Xu, Z. Wang, H. Guo, K. Zhang (2014). *Nano Energy*, **6**, pp. 19–26.
17. J. Wan, Q. Zhang, X. Li, B. Zhanga, L. Mai, K. Zhang (2015). *Nano Energy*, **12**, pp. 437–446.
18. Z. Li, Y. Xin, Z. Zhang, H. Wu, P. Wang (2015). *Sci. Rep.*, **5**, p. 10617.
19. C. Yuan, L. Yang, L. Hou, L. Shen, X. Zhang, X. W. Lou (2012). *Energy Environ. Sci.*, **5**, pp. 7883–7887.
20. J. Tang, Y. Ge, J. Shen, M. Ye (2016). *Chem. Commun.*, **52**, pp. 1509–1512.
21. Z. Gu, R. Wang, H. Nan, B. Geng, X. Zhang (2015). *J. Mater. Chem. A*, **3**, pp. 14578–14584.
22. S. Yang, P. Yan, Y. Li, K. Cheng, K. Ye, C. Zhang, D. Cao, G. Wang, Q. Li (2015). *RSC Adv.*, **5**, pp. 87521–87527.

Chapter 6

Hybrid Reduced–Graphene Oxide/$MnSe_2$ Cubes: A New Electrode Material for Supercapacitors

In this chapter, the integration of 2D graphene nanosheets and layered transition metal dichalcogenides has been recognized as one of the most extensive strategies for obtaining promising electrode materials for energy storage devices. Cubic manganese diselenide ($MnSe_2$) and rGO-decorated $MnSe_2$ (G-$MnSe_2$) hybrid materials were synthesized via a facile hydrothermal method.

On the other hand, the metallic selenium impurity is considered a major unwanted by-product in this method. An effective means to remove such bulk chalcogenide is a key challenge. For the synthesis of a G-$MnSe_2$ hybrid material, a swift strategy was adopted by mixing the graphene oxide with manganese and selenium precursors. Surprisingly, the final G-$MnSe_2$ product contained a negligible amount of selenium impurity. The $MnSe_2$ and G-$MnSe_2$ hybrid materials were characterized in detail. For the first time, the electrochemical energy storage behavior of manganese selenide-based materials was assessed for supercapacitor applications. The specific capacitance of the $MnSe_2$ electrode was approximately

Solar Power and Energy Storage Systems
Hee-Je Kim

ISBN 978-981-4800-13-6 (Hardcover), 978-0-429-45877-4 (eBook)
www.panstanford.com

57.8 mF cm^{-2}, whereas the hybrid G-$MnSe_2$ electrode showed a much higher specific capacitance, of 93.3 mF cm^{-2}, at a scan rate of 1 mV s^{-1}. A symmetrical cell made from the G-$MnSe_2$ hybrid material shows excellent long-term stability for 4500 cycles and approximately 106% retention of its initial capacitance, which is impressive compared to the long cycle life of $MnSe_2$-based symmetrical cells (80% capacitance retention at the 4500 cycles).

6.1 Overall Survey

The current development of renewable energy conversion systems requires efficient electrochemical energy storage devices.[1] Among the various electrochemical energy storage devices, the supercapacitor is considered one of the most promising energy storage devices owing to its excellent electrochemical properties, such as high power density, rapid charging rate, and long-term cycling stability.[2] These features have been used in a variety of applications, such as hybrid electric vehicles, digital communication devices, power electronics, backup power for windmills, rubber tire gantry cranes, military applications, and smart grid applications.[3] Supercapacitors consist of four main parts: positive electrode, negative electrode, separator, and electrolyte. The key performance of the supercapacitor is based on the energy storage capacity of the electrode materials and the stability of the electrolyte for a wide range of operating voltages.[4] To date, various electrode materials have been investigated for supercapacitor applications,[5] such as carbon-based nanomaterials,[6] conducting polymers (polyaniline, polypyrrole, and polythiophene),[7] transition metal oxides (TMOs),[8] transition metal nitrides (TMNs),[9] and transition metal dichalcogenides (TMDs).[10] Carbon-based materials are the best-known double-layer capacitance electrodes, but the fundamental shortcoming is their low specific capacitance and inferior energy density. Thus far, various conducting polymer–based electrode materials have been investigated for supercapacitor applications. Although conducting polymer–based electrodes show a high specific capacitance, their cycling stability remains for a few hundred cycles only.[11] TMOs have been investigated widely for supercapacitor applications and show a good specific capacitance; however, the electrical conductivity of TMOs is very low, which increases the overall resistance of the device.[12] The synthesis of TMNs is very difficult due to the formation of N≡N bonds and its instability in air and humid atmosphere, which can affect the physicochemical properties and, in turn, the energy storage performance of the electrode.

In the search for suitable pseudocapacitive electrode materials, TMDs are considered as one of the propitious candidates, in which sulfur, selenium, and tellurium come from the same main group as oxygen and can be reacted with various transition metal group elements.[13] Recently, TMDs have attracted considerable attention in the supercapacitor field because of their good electrical conductivity, 2D sheet-like layered crystal structure, high surface area, and various redox states of transition metal ions. Selenium is a nonmetal with intermediate chemical activity and physical properties between that of sulfur and tellurium. Transition metal diselenides have recently been proposed as interesting candidates for energy storage devices, such as batteries (Li-ion and Na-ion batteries) and supercapacitors.[14] Recently, molybdenum diselenide ($MoSe_2$)-based materials have been assessed for supercapacitor applications.[15] The inspiring results of materials based on $MoSe_2$, $NiSe_2$,[16] $CuSe_2$,[17] and WSe_2[18] prompted this study to synthesize manganese dichalcogenide–based materials, which has never been investigated for supercapacitor applications.[19] Although $MnSe_2$ possesses a crystal structure and other intrinsic chemical properties similar to other metal TMDs, manganese is an abundant element in the earth's atmosphere, with low toxicity and relatively low cost.[20] On the other hand, the synthesis of pure $MnSe_2$ by the hydrothermal method is crucial due to the formation of undesirable selenium impurities.[21] The removal of selenium impurities from the final $MnSe_2$ product has not been investigated by many researchers. Zheng et al. attempted to remove the selenium impurity by baking their final $MnSe_2$ product at 503 K for 15 min.[22] When the same eviction method was adopted, no selenium impurity peaks were observed in the X-ray diffraction (XRD) pattern; in contrast, strong SeO_2 peaks could be observed by X-ray photoelectron spectroscopy (XPS). Zheng et al.'s calcination method did not remove the selenium impurity completely; instead, selenium might have been converted to an amorphous phase that could not be observed by XRD. In general, pristine metal dichalcogenides have some intrinsic demerits, such as chemical instability, low specific capacitance and inferior rate capability, and poor cycle life, that limit their electrochemical energy storage performance on a device scale. These drawbacks of pristine metal dichalcogenides could be overcome by combining them with highly conducting carbon-based materials to form a hybrid structure. Various carbon and carbon-based electrode materials have been used for supercapacitor applications (carbon nanotubes, graphite, activated carbon and graphene, *etc.*). Among them, graphene is the most interesting material owing to its high surface area and excellent electrical and chemical properties.[23]

In this study, a facile strategy was reported to diminish the concentration of selenium impurities in the reduced graphene oxide (rGO)-decorated $MnSe_2$ (G-$MnSe_2$) material by an in situ reaction of GO with a manganese and selenium source.[24] The one-step hydrothermal synthesis led to the formation of the G-$MnSe_2$ hybrid material. More interestingly, when the GO was mixed with the manganese and selenium precursors in this method, the final product, G-$MnSe_2$, had an insignificant amount of selenium impurity without the need for an external purification or calcination procedure. The as-synthesized G-$MnSe_2$ hybrid material showed much better electro chemical charge storage properties than that of the rGO-free $MnSe_2$ material.[25] To the best of our knowledge, this is the first study of $MnSe_2$ and G-$MnSe_2$-based materials for supercapacitor applications. The G-$MnSe_2$-based symmetric capacitor showed a very high areal-specific capacitance (93.25 mF cm^{-2}), an energy density of 1.84 mWh cm^{-2}, and promising long-term cycling stability over 4500 cycles.

6.2 Materials and Methods

A schematic representation of the synthesis of a cubic G-$MnSe_2$ hybrid material is shown in Fig. 6.1. The crystal structure and phase purity of the as-prepared electrode materials were characterized by XRD.

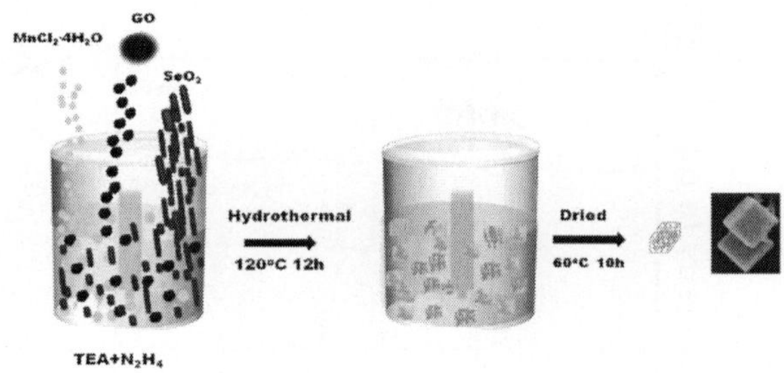

Figure 6.1 The schematic representation of the synthesis of a cubic G-$MnSe_2$ hybrid material.

Figure 6.2a shows the XRD pattern of GO, rGO, $MnSe_2$, and G-$MnSe_2$. The intense diffraction peak of GO was found at 2θ = 11.12° (002 plane), which corresponds to an interlayer distance (d-spacing) of 0.79 nm. Obviously, the interlayer distance of GO is higher than the interlayer distance of pristine graphite (d = 0.33 nm). The oxidation of graphite leads to the formation of oxygen-containing functional groups in the sp^2 domain. In addition, functionalization leads to the formation of structural defects and new functional groups are introduced in the sp^3 orbital, which leads to a higher d-spacing in the GO crystal structure. After the hydrothermal reaction, GO was reduced to rGO and the (002) diffraction peak appeared at 23.58° 2θ, which is due to the removal of oxygen-containing functional groups from GO. This confirms the efficient reduction of GO to rGO. From Fig. 6.2a, the sharp and intense diffraction peaks indicate the formation of highly crystalline $MnSe_2$. The characteristic peaks of $MnSe_2$ could be indexed to the cubic phase with a lattice constant of a = 6.4170 Å (JCPDS 73-1525).[21] In addition to the $MnSe^2$ peaks, small peaks appear at 29.5°, 43.6°, and 45.21° 2θ due to the existence of residual selenium impurity.[29] These residual selenium peaks disappeared during calcination of the sample at 230°C for 15 min. As the melting point of selenium is approximately 221°C, the evaporation of selenium is difficult at this temperature for a short calcination time (see Supporting Information, Fig. S1).[30] This suggests that heating the sample leads to the formation of poorly crystalline impurity peaks, which cannot be detected by XRD. To further understand the effects of the heating

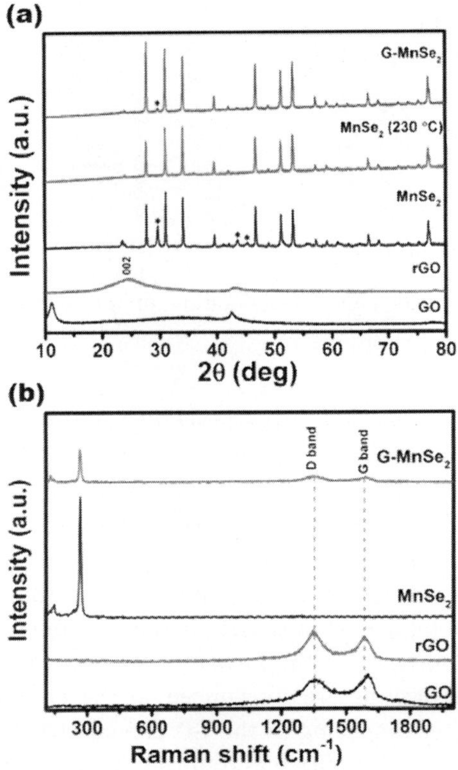

Figure 6.2 (a) XRD patterns of GO, rGO, $MnSe_2$, $MnSe_2$ (230°C), and G-$MnSe_2$. (b) Raman spectrum of GO, rGO, $MnSe_2$, and G-$MnSe_2$.

process, the calcined sample was examined by XPS (described later). The as-synthesized G-$MnSe_2$ hybrid material exhibited an XRD pattern similar to the cubic $MnSe_2$ phase.

Interestingly, the selenium impurity peaks were reduced significantly (almost to a negligible amount) without further calcination. The addition of GO refined the $MnSe_2$ crystal structure in the

G-$MnSe_2$ hybrid material, which is the key finding of this study. The GO sheets increased the segregation of selenium ions (Se^{2-}) in the precursor solution. During the hydrothermal reaction, the well-dispersed Se^{2-} ions reacted with Mn^{2+} ions to form the pure $MnSe_2$ in the G-$MnSe_2$ hybrid material. This implies that addition of GO is profoundly responsible for the segregation of Se^{2-} ions and, in turn, reduces the Se-Se interaction, thereby preventing the formation of selenium metal impurities.[31] Unfortunately, no obvious peaks related to rGO were observed in the G-$MnSe_2$ sample, which might be due to the small amount of carbon and relatively low diffraction intensity of rGO in the G-$MnSe_2$ hybrid material.[32] Therefore, the diffraction peaks of rGO become much weaker after forming the G-$MnSe_2$ hybrid material. Nevertheless, the presence of rGO in the G-$MnSe_2$ hybrid material was confirmed by XPS. The structural information of the GO, rGO, $MnSe_2$, and G-$MnSe_2$ hybrid material was further confirmed by Raman spectra, as shown in Fig. 6.2b. The Raman spectrum of GO showed two high-intensity peaks, appearing at wavelengths of 1356 cm^{-1} and 1600 cm^{-1}, which are attributed to the characteristic D and G bands, respectively. The rGO reveals characteristic D and G bands at the corresponding wavelengths of 1347 cm^{-1} and 1581 cm^{-1}, respectively. Furthermore, there was an obvious increase in the intensity ratio of ID/IG in the case of rGO compared to the GO spectrum, which indicates the deoxygenation in rGO. The higher value of the ID/IG ratio confirms the formation of rGO from GO.[33] From Fig. 6.2b, the typical $MnSe_2$ peaks were observed at 267 cm^{-1} and 147 cm^{-1}, which correspond to the out-of-plane Ag mode and the in-plane Eg mode of $MnSe_2$.[34] The Raman spectrum of the G-$MnSe_2$ material shows the combination of typical $MnSe_2$ and rGO peaks, which confirms the formation of a hybrid material. Field-emission scanning electron microscopy (FE-SEM) was used to characterize the surface morphology and particle size.

Figure 6.3 shows FE-SEM images of hydrothermally synthesized $MnSe_2$ and G-$MnSe_2$ cubes at different magnifications. Figure 6.3a shows uneven $MnSe_2$ cubes with scratches on the surface. The average dimension of the $MnSe_2$ cubes was around 6 μm, which is depicted in the high-magnification image (Fig. 6.3b). In the case of the G-$MnSe_2$ hybrid material, however, rGO nanoneedles grew on the surface of the cubes, which was clearly observed from the high-magnification FE-SEM images (Figs. 6.3c and 6.3d). The formation of GO, rGO, $MnSe_2$, and G-$MnSe_2$ hybrid materials was confirmed by XPS. From the survey spectra (see Supporting Information, Fig. S2), the C 1s, Mn 2p, Se 3d, and N 1s oxidation states of the elements can be identified clearly at 284.6, 641.90, 54.88, and 399.4 eV,

Figure 6.3 FE-SEM images of (a, b) a cubic $MnSe_2$ material and (c, d) G-$MnSe_2$ hybrid materials of the as-synthesized materials.

respectively. For GO, the various carbon peaks appear at 284.53, 286.54, and 288.43 eV, which indicate the three different functional groups formed with carbon (see Supporting Information, Fig. S3). The peak at 284.53 eV was assigned to the nonoxygenated C–C bond, which is assigned to sp^2-hybridized carbon atoms. The other peak, at 286.54 eV, was attributed to the C–O bond, and the last one corresponds to the carboxylate carbon, C=O bond.[35] These binding energies are consistent with the reported values of GO. For the GO material, the atomic ratios of O 1s/C 1s were 0.56. Figure S3 (Supporting Information) shows the C 1s spectra of the rGO material. After reduction, the intensity of the C=O peak (285.33 eV) decreased significantly and the C–C peak (at 284.58 eV) of the sp^2-hybridized carbon became predominant in the rGO material. This confirms that a considerable amount of deoxygenation occurs in rGO, leading to the formation fairly reduced rGO. To further confirm this reduction process, the O 1s/C 1s atomic ratio was calculated on the basis of the peak intensity and the value was significantly reduced to 0.10.

Similarly, the G-$MnSe_2$ hybrid material showed a carbon peak at 284.60 eV, as well as a lower amount of oxygen-functionalized

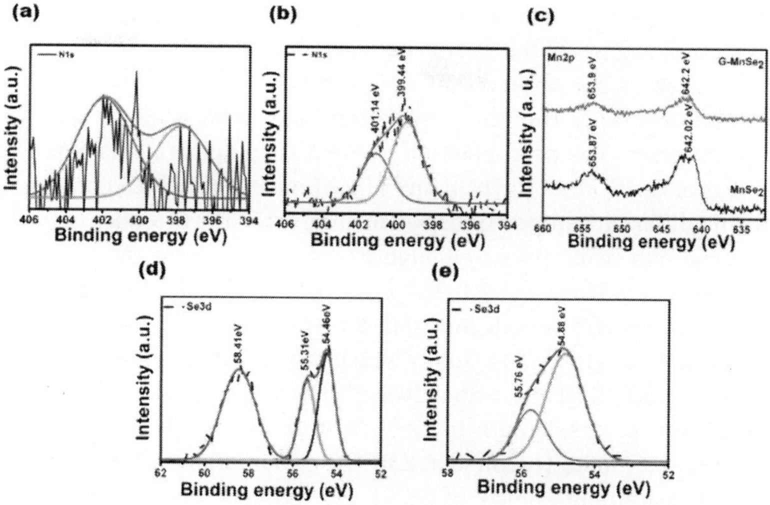

Figure 6.4 High-resolution XPS spectra for (a) N 1s transition of rGO, (b) N 1s transition of G-$MnSe_2$. (c) Mn 2p peaks of $MnSe_2$ and G-$MnSe_2$, (d) Se 3d transition of $MnSe_2$, and (e) Se 3d transition of G-$MnSe_2$.

carbon components than that of GO, which indicates the presence of rGO in the G-$MnSe_2$ hybrid material. For the rGO and G-$MnSe_2$ hybrid materials (Figs. 6.4a and 6.4b), the N 1s peak was also observed at approximately 400 eV, which confirmed the presence of nitrogen in these materials. The two deconvoluted peaks were detected in this region using a Gaussian fit. The peak positions at 399.4 and 401.1 eV correspond to the pyridinic and pyrrolic nitrogen species, respectively, in the rGO and G-$MnSe_2$ samples. The presence of N 1s peaks might be due to the interaction between GO and hydrazine hydrate ($N_2H_4 \cdot H_2O$). On the other hand, for the $MnSe_2$ sample, the N 1s peak was not observed, which confirms that the nitrogen is doped into the rGO but not into the metal selenide. Doping of nitrogen into the rGO material further enhances the electrical conductivity of the material.[36] High-resolution spectra of Mn and Se were performed in their core regions. Figure 6.4c shows the high-resolution Mn 2p spectra of the $MnSe_2$ and G-$MnSe_2$ samples. The two distinct peaks at 642.2 and 653.9 eV correspond to the 2p electron spin-orbit components of Mn $2p^{3/2}$ and Mn $2p^{1/2}$, respectively. The energy difference of $\Delta 2p$ was 11.7 eV, which confirms the presence of Mn^{4+} in the sample. A good correlation was observed with the literature value for the Mn^{4+} state.[37] No Mn^{2+} peaks were observed at 640 $\approx$ 641 eV, which confirms the formation of only the Mn^{4+} oxidation state with a d3 electronic configuration. The peak position of the Mn 2p spectra was similar to that of the G-$MnSe_2$ hybrid and $MnSe_2$ materials, which confirms that no different compositions had formed while the G-$MnSe_2$ hybrid material was being prepared. Figures 6.4d and 6.4e present the Se 3d spectra of $MnSe_2$ and G-$MnSe_2$ hybrid materials, respectively. The Se 3d spectra was deconvoluted further into two well-defined peaks, such as $3d^{5/2}$ and $3d^{3/2}$, which appear at binding energies of 54.46 and 55.31 eV, respectively.[38] These two peaks confirmed the presence of selenium in Se^{2-} form in $MnSe_2$. In addition to these peaks, $MnSe_2$ (calcined at 230°C for 15 min.) has an additional peak at a binding energy of 58.41 eV,[38] which was attributed to the formation of SeO_2 during calcination. Interestingly, such oxide peaks were not observed in the G-$MnSe_2$ hybrid material. Zheng et al.'s[22] calcination method was used to remove the small quantity of Se impurities from $MnSe_2$. XRD indicated the removal of selenium impurities. In contrast to the XRD observations, XPS confirmed the existence of selenium in the form of an oxide (SeO_2). Therefore, while the sample was being heated to the melting point of selenium (approximately 221°C), the corresponding selenium impurity might have converted to either an amorphous phase of SeO_2 or the presence of trace amounts, maybe undetectable by XRD. In the G-$MnSe_2$ hybrid, however, there was no noticeable amount of selenium impurities observed from the XRD and XPS results. This is mainly due to the addition of GO (rGO) sheets, which suppressed the rate of the reaction between Mn and H_2Se (intermediate formed) gas, leading to the formation of high-purity $MnSe_2$ decorated with rGO nanoneedles.[32] The amount of graphene present in the G-$MnSe_2$ hybrid material was determined by thermogravimetric analysis (TGA), as illustrated in Fig. S4. The electrochemical performance of the as-synthesized electrode materials ($MnSe_2$ and G-$MnSe_2$) was evaluated using the two-electrode symmetrical cell configuration with an aqueous 0.5 M H_2SO_4 electrolyte. The specific capacitance of the cell and the corresponding specific capacitance of a single electrode were determined by a cyclic voltammetry (CV) and CD.

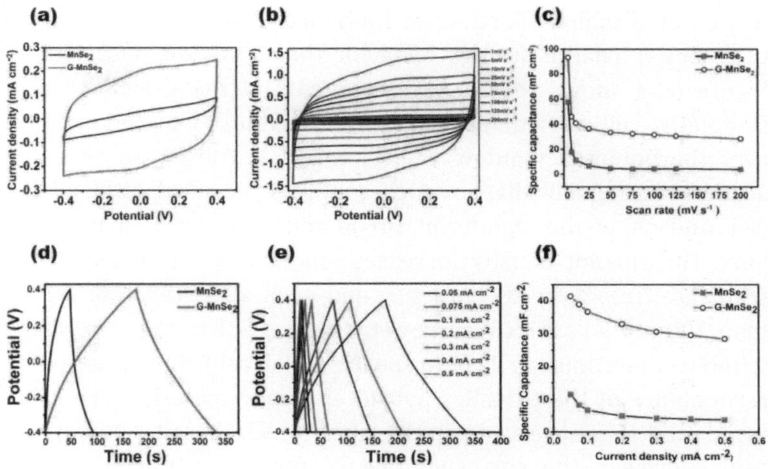

Figure 6.5 (a) CV curves of cubic $MnSe_2$ and G-$MnSe_2$ symmetric cells measured at a scan rate of 25 mV s^{-1}. (b) CV curves of G-$MnSe_2$-based symmetric cells measured at various scan rates, ranging from 1 to 200 mV s^{-1}. (c) The areal capacitance of the $MnSe_2$ and G-$MnSe_2$ hybrid electrodes measured at different scan rates. (d) GCD curves of $MnSe_2$- and G-$MnSe_2$-based symmetric cells measured at a current density of 0.3 mA cm^{-2}. (e) GCD curves of G-$MnSe_2$-based symmetric cells measured at various current densities. (f) The areal-specific capacitance of the $MnSe_2$ and G-$MnSe_2$ electrodes measured at different current densities.

Figure 6.5a compares the CV curves of $MnSe_2$- and G-$MnSe_2$-based symmetric cells measured at a scan rate of 25 mV s^{-1}. Both CV curves showed a quasi-rectangular shape, which confirmed the ideal electric double-layer capacitance (EDLC) behavior. In addition to the quasi-rectangular shape, the CV of the $MnSe_2$ cell showed a pair of broad redox peaks, which correspond to the electrochemical redox reaction of Mn^{4+} ions. The chemical reactions of the faradaic and nonfaradaic energy storage processes are given in Eqs. 6.1 and 6.2, respectively.[40]

$$(\mathrm{MnSe_2}) + \mathrm{H}^+ + \mathrm{e}^- \mathrm{MnSe} - \mathrm{SeH}^+ \tag{6.1}$$

$$(\mathrm{MnSe_2})\mathrm{surface} + \mathrm{H}^+ + \mathrm{e}^- (\mathrm{MnSe_2} - \mathrm{H})\mathrm{surface} \tag{6.2}$$

Interestingly, the CV of the G-$MnSe_2$-based symmetric cell showed a much greater integrated area as well as a pair of more pronounced broad redox peaks. This suggests that G-$MnSe_2$

possesses a higher electrochemically active surface area and, in turn, better charge storage behavior than those of pure $MnSe_2$. Figure 6.5b shows the CV measurements of the G-$MnSe_2$-based symmetric cell carried out at various scan rates (1 to 200 mV s^{-1}) over the potential window of −0.4 to 0.4 V. All CV curves show a quasi-rectangular shape, which confirms the ideal EDLC of the cell, and show the significant broad redox peaks of manganese ions. The current density increases gradually with an increasing scan rate from 1 to 200 mV s^{-1}, and even at a scan rate higher than 200 mV s^{-1}, the CV curve retains its quasi-rectangular shape without compromising its redox peaks, which confirms the excellent reversibility of the G-$MnSe_2$ hybrid electrode material. The same trend was also observed for the pure $MnSe_2$–based symmetrical cells; however, the corresponding current density and specific capacitance were much lower than those for the G-$MnSe_2$-based cells (see Supporting Information, Fig. S5). At a low scan rate of 1 mV s^{-1}, the corresponding specific capacitance of the $MnSe_2$-based symmetrical cell was 28.9 mF cm^{-2} and the corresponding single electrode–specific capacitance was 57.8 mF cm^{-2}. Compared to the $MnSe_2$ material, the G-$MnSe_2$ hybrid material showed a higher specific capacitance—46.6 mF cm^{-2} for a cell and 93.3 mF cm^{-2} for a single electrode. The CE value obtained in this study using the two-electrode configuration is still superior to that of the amorphous MoS_x material (83.9 mF cm^{-2} at 1 mV s^{-1}).[28] When the scan rate was increased to 10 mV s^{-1}, the CE value decreased to 40.1 mF cm^{-2}. On the other hand, the CE value obtained in the present work was 3 times higher than that of the paintable MoS_2 (8 mF cm^{-2}) and 10 times higher than that of the exfoliated MoS_2-based symmetric devices (2 mF cm^{-2}).[40] The excellent electrochemical performance of the G-$MnSe_2$-based hybrid material may be due to the following reasons: (i) $MnSe_2$ possesses a 2D layered crystal structure, which paves the way for the intercalation of cations (H^+ ions from electrolyte) into the crystal matrix, (ii) TMDs have high electrical conductivity (when compared to oxides), which reduces the charge-transfer resistance of the electrode material, (iii) the synergetic effect of rGO nanoneedles decorating the $MnSe_2$ cubes enhanced the charge storage capacity of the material via both faradaic (redox) and nonfaradaic (EDLC) energy storage processes,

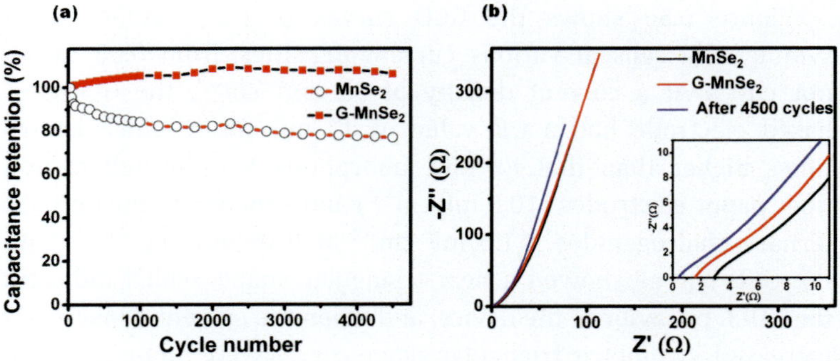

Figure 6.6 (a) Long-term stability of $MnSe_2$- and G-$MnSe_2$-based symmetric cells measured at a constant current density of 0.3 mA cm^{-2} for 4500 cycles. (b) Electrochemical impedance spectra of $MnSe_2$-based symmetric cells (black line), G-$MnSe_2$-based symmetric cells measured before long-term stability test (1st cycle, red line), and G-$MnSe_2$-based symmetric cells measured after long-term stability test (4500 cycles, blue line). Although both curves showed near triangular shapes with nonlinear behavior, the discharge time and specific capacitance of the G-$MnSe_2$ electrode were three times higher than those of the $MnSe_2$ electrode.

and (iv) the presence of highly conducting rGO material and the effect of nitrogen doping of rGO sheets (in G-$MnSe_2$ hybrid) further improve the electrical conductivity and, in turn, reduce the overall resistance of the device.[41] The relationship between the specific capacitance of a single electrode and the scan rate is shown in Fig. 6.5c. The specific capacitance of a single electrode decreased gradually with an increasing scan rate, which might be due to the mass transport limitation of electrolyte ions from the bulk electrolyte to the electrode surface as well as due to the insufficient time available for the redox reaction inside the bulk part of electrode materials.[42, 43] To further evaluate the electrochemical performance of $MnSe_2$- and G-$MnSe_2$-based electrode materials, galvanostatic charge-discharge (GCD) measurements were carried out for symmetrical cells at a constant current density of 0.3 mA cm^{-2} (see Fig. 6.5d). For a $MnSe_2$-based cell, the specific capacitance of the single electrode was 11.6 mF cm^{-2}, and for the G-$MnSe_2$-based material, the specific capacitance of a single electrode reached 41.5 mF cm^{-2}.

Figure 6.5e shows the GCD curves of the G-$MnSe_2$-based symmetrical cells at various current densities, from 0.05 to 0.5 mA cm^{-2}. At a current density of 0.2 mA cm^{-2}, the G-$MnSe_2$-based electrode had a CE value of 33 mF cm^{-2}, which is two times higher than that for the amorphous MoS_x-coated carbon fiber paper electrodes (10.5 mF cm^{-2}) and superior to the $CoNi_2S_4$ ternary chalcogenides (11.6 mF cm^{-2} at 0.16 mA cm^{-2}).[28,44] All the GCD curves showed a near-triangular shape, which indicates the EDLC behavior of the device, and when the current density was increased, symmetric triangular shape curves were retained, which indicates ideal capacitive behavior. On the other hand, nonlinear behavior was only observed at a low current density, which might be due to the sufficient time available for the redox reaction to occur on the inner part of the electrode material. Therefore, a higher pseudocapacitance contribution was observed with the low-current-density GCD measurements. Figure 6.5f shows the relationship between the specific capacitance of a single electrode (calculated from $MnSe_2$ and G-$MnSe_2$ symmetrical cells) at various current densities (0.05–0.5 mA cm^{-2}). The calculated specific capacitance decreased with increasing current density. This is because at higher current densities, there was insufficient time available for the efficient diffusion of H^+ ions into the bulk part of the electrode, which limits the occurrence of redox reactions to the outer surface of the electrode. With the low-current-density GCD measurements, the electrolyte ions have sufficient time to penetrate into the interior (bulk) part of the electrode surface, resulting in higher capacitance.[12] Figure S6 (see Supporting Information) presents a Ragone plot of the G-$MnSe_2$-based symmetric cell. The energy density and power density were calculated using Eqs. 6.3 and 6.4. The G-$MnSe_2$-based symmetrical cell achieved the highest energy density of 1.84 mWh cm^{-2} at a corresponding power density of 19.9 mW cm^{-2}. At the highest power density, of 200 mW cm^{-2}, 68% of the energy density was still retained (1.26 mWh cm^{-2}), which highlights the good performance of the cell. For commercial applications of supercapacitors, a long cycle life is an important parameter to recognize the stability of the electrode materials. The cycling stability of the $MnSe_2$- and G-$MnSe_2$-based symmetric cells was measured at a high current density of 0.3 mA

cm^{-2} for 4500 cycles. Figure 6.6a shows the capacitance retention versus cycle number of both $MnSe_2$- and G-$MnSe_2$-based cells. For a pure $MnSe_2$-based symmetric cell, the specific capacitance and capacitance retention decreased abruptly for the first few hundred cycles and then decreased gradually until 4500 cycles. Right after the 4500 cycles were completed, around 75% of its initial capacitance value was retained. In contrast to this observation, the G-$MnSe_2$-based cell showed a gradual increasing trend for 2000 cycles and then the specific capacitance was almost constant until the end. Right after the 4500 cycles were completed, the hybrid electrode showed retention of approximately 106% of its initial capacitance, which confirmed the excellent long-term stability of this hybrid electrode material that suits real-time applications. An increasing trend of capacitance retention during the long-term stability test was also observed. For example, graphene-based materials, amorphous MoS_x coated on carbon fiber paper, and graphene and MoS_2 hybrid materials showed an increasing trend of capacitance retention during the continuous charge-discharge process. The authors explained that the increasing capacitance trend might be due to the "electroactivation" of the electrode materials during the repeated charge-discharge cycling process. During the electroactivation process, the electrolyte ions might have intercalated into the 2D layered materials, thereby either exfoliating them or broadening the electrochemically active surface area for the enhanced charge storage behavior.[45c,d] To understand the internal resistance and fundamental capacitive behavior of the fabricated cell, electrochemical impedance spectroscopy (EIS) was performed on the $MnSe_2$- and G-$MnSe_2$-based symmetric cells.[46] Figure 6.6b shows the electrochemical impedance spectra of the $MnSe_2$-based symmetric cells and G-$MnSe_2$-based symmetric cells at the initial stage and after 4500 charge-discharge cycles. All the Nyquist plots showed a semicircular arc in the high-frequency region and a straight line in the low-frequency region. The x-axis intercept of the semicircle observed in the high-frequency region represents the electrochemical series resistance (ESR) of the symmetric cells. The semicircle observed in the high-medium frequency region was attributed to the charge transfer resistance (R_{ct}) at the electrode and electrolyte interface. The low-frequency region showed almost a vertical straight-line behavior, which represents the ideal capacitive behavior of the electrode material. The ESR value of the $MnSe_2$-based symmetrical cell was approximately 2.92 Ω. On the other hand, the hybrid G-$MnSe_2$ showed a lower ESR value—of approximately 1.60 Ω. The presence of rGO nanoneedles on the $MnSe_2$ cubes increases the electrical conductivity of the electrode material, resulting in a reduced ESR value. The R_{ct} values of $MnSe_2$ and G-$MnSe_2$ were observed from the high-frequency region of the impedance spectra (see inset of Fig. 6.6b). The R_{ct} values of the $MnSe_2$- and G-$MnSe_2$-hybrid electrode were 1.8 Ω and 1.2 Ω. Compared to $MnSe_2$, the G-$MnSe_2$ hybrid electrode showed lower R_{ct} values, indicating the facile charge transfer and higher electrical conductivity. After 4500 cycles, the R_{ct} value of the G-$MnSe_2$ hybrid material electrode further decreased from 1.2 Ω to 0.6 Ω, which confirmed the synergistic effects of the rGO and $MnSe_2$ materials, its good electrical conductivity, and facile charge-transfer kinetics. The electrochemical impedance spectrum of the G-$MnSe_2$-based symmetric cells was tested after the long-term stability test (4500 cycles). Interestingly, the G-$MnSe_2$-based cells after the long-term stability test showed much lower ESR and R_{ct} values than those of the initial measurement. These reduced ESR and R_{ct} values may be due to the electroactivation process and/or sufficient soaking time of the electrode materials in the aqueous electrolyte, which activates the binder during the repeated charge-discharge process. This confirms that during the long-term cycling process, the stability of the electrode material increases further; therefore, increasing specific capacitance retention is observed from the long-term stability test (approximately 106% capacitance retention).

6.3 Conclusions

Cubic $MnSe_2$ and hybrid G-$MnSe_2$ electrode materials were synthesized via a facile hydrothermal method. The addition of GO helped to form a highly conducting rGO-decorated $MnSe_2$ hybrid material. Interestingly, the G-$MnSe_2$ hybrid material contained negligible selenium impurities (without additional purification/calcination process) compared to the $MnSe_2$ material. The hybrid G-$MnSe_2$-based symmetric cells exhibited a high areal capacitance (of 46.6 mF cm^{-2}) and energy density of 1.84 mWh cm^{-2} as well as enhanced cycling stability of 106% over 4500 cycles. The calculated areal capacitance of the G-$MnSe_2$ electrode was 93.3 mF cm^{-2}, which is much higher than that of the $MnSe_2$-based electrodes (57.8 mF cm^{-2}). The increased specific capacitance, high energy density, and very low impedance of the G-$MnSe_2$-based symmetric cells confirmed the synergistic effect of rGO and $MnSe_2$ materials in this hybrid electrode. Decreasing the particle size of the G-$MnSe_2$ cubes to the nanometer scale is expected to further increase the electrochemically active surface area and specific capacitance. Studies in this area are currently underway.

Appendix

A6.1 Experimental Section

A6.1.1 Synthesis of rGO, Pristine $MnSe_2$, and G-$MnSe_2$ Materials

Analytical-grade reagents and chemicals were used without further purification. GO was synthesized from graphite powder using a modified Hummer's method.[26] A 0.3548 g sample of GO was sonicated in 50 mL deionized (DI) water for 30 min. Subsequently, 10 mL of hydrazine hydrate was added slowly to the GO mixture with constant stirring. Approximately 30 min. later, the precursor solution was transferred to an 80 mL capacity Teflon-lined container that was transferred immediately into a stainless-steel (SS) autoclave and then tightly sealed. The autoclave was heated in a hydrothermal instrument at 120°C for 12 h. After the reaction was completed, the autoclave was cooled naturally to room temperature.

The final product was washed several times with DI water and ethanol to remove the excess hydrazine hydrate from the rGO material. Finally, it was dried in an electric oven at 60°C for 10 h. The G-$MnSe_2$ hybrid material was prepared as follows: A

0.3548 g sample of GO was dispersed in 20 mL of DI water and sonicated for 30 min. To another beaker, an exactly 1:2 molar ratio of manganese (II) chloride tetrahydrate ($MnCl_2 \cdot 4H_2O$, purity $\geq$ 98.5%, Sigma Aldrich) and selenium dioxide powder (SeO_2, purity $\geq$ 99.5%, SAMCHUN) were mixed in 25 mL of DI water under vigorous stirring for 15 min. To this mixture, an appropriate amount of triethanolamine—(C_2H_5)3N, concentration $\geq$ 99.0%, Sigma Aldrich—was added and sonicated for 10 min. Triethanolamine acted as a surface stabilizer to reduce the aggregation for $MnSe_2$ particles and increase the chemical bonding between Mn and Se. Hence, a more stabilized cube-like structure hydrate ($N_2H_4 \cdot H_2O$, concentration 50%–60%, Sigma Aldrich) was poured into the above precursor solution and stirred vigorously for a further 30 min. The GO dispersion (0.3548 g of GO in 20 mL of DI water) was added dropwise to the above mixture with continuous stirring for 30 min. to ensure a uniform dispersion of GO in the precursor solution. The as-prepared precursor solution was transferred to a Teflon-lined container. The Teflon container was transferred to a SS autoclave and then tightly sealed. The autoclave was heated in a hydrothermal instrument, and the reaction temperature was maintained at 120°C for 12 h. After the reaction was completed, the autoclave was allowed to cool naturally to room temperature. The resulting product, the G-$MnSe_2$ hybrid material, was washed several times with DI water and anhydrous ethanol to remove any unwanted residues and then dried overnight in an electric oven at 60°C. For comparative purpose, pure $MnSe_2$ was also synthesized using a similar method without the addition of GO. An attempt was made to eliminate selenium impurities by calcining the sample at 230°C for 15 min. in air according to the methodology reported by Zheng et al.[22]

A6.1.2 Material Characterization

The crystal structure and lattice parameters of the as-prepared samples were characterized by XRD using Cu-Kα radiation (λ = 1.5418 A) over the 2θ range, 10° and 80°, at a scan rate of 2° min^{-1}. The surface morphology and elemental composition of the materials were examined by FE-SEM combined with energy-dispersive X-ray spectroscopy (Japan Hitachi: Model number S-4200). The chemical composition of the sample was examined by XPS using Al-Kα radiation (hv = 1486. 6 eV, Thermo Fisher Scientific, UK). Raman spectroscopy (WITec) was used with 532 nm laser excitation, after calibrating the Raman shift with a silicon reference at 521 cm^{-1}. The thermal stability of the G-$MnSe_2$ hybrid materials was analyzed by TGA (Q50 V6.2, Build 187, TA instruments, US).

A6.1.3 Electrochemical Measurements

To evaluate the electrochemical performance of the as-synthesized materials, the working electrode was prepared by first mixing 80% of the active materials with 10% of carbon black and 10% of polyvinylidene fluoride as the binder. The components were ground using a mortar and pestle, and a few drops of anhydrous 1-methyl 2-pyrrolidone solvent were added to form a homogenous paste. The geometrical area of the electrode was fixed to 1 cm^2. Initially, the SS plates were cleaned sequentially with a 5 N H_2SO_4 solution for 30 s, DI water, and acetone and then dried at 100°C in an electric oven for 15 min. The SS substrate was coated with the homogeneously mixed paste using a brush. The electrode was then dried at 60°C for 12 h in a vacuum oven to ensure complete evaporation of the solvents. One electrode was placed inside the standard two-electrode test cell setup. A small piece of Whatman filter paper was placed on it as a separator. To this, a few drops of a 0.5 M H_2SO_4 electrolyte were added and another electrode was placed over the separator. The split test cell was assembled and left for a few hours to ensure uniform saturation of the electrodes in the electrolyte solution prior to the electrochemical measurements. The electrochemical properties were measured using a SP-150 biological instrument. CV was performed at different scan rates (1 to 125 mV s^{-1}) in the potential window of –0.4 to 0.4 V. GCD curves were recorded at various current densities (0.05 to 0.5 mA cm^{-2}) in a potential window of –0.4 to 0.4 V. EIS was performed over the frequency range of 1 Hz to 100 kHz with an AC amplitude of 10 mV at 0 V DC bias.

References

1. (a) A. Burke (2000). *J. Power Sources*, **91**, p. 37; (b) K. W. Ko, M. Lee, S. S. Sekhon, S. K. Balasingam, C. H. Han, Y. Jun (2013). *ChemSusChem*, **6**, p. 2117.
2. R. Kotz, M. Carlen (2000). *Electrochim. Acta*, **45**, p. 2483.
3. A. L. M. Reddy, S. R. Gowda, M. M. Shaijumon, P. M. Ajayan (2012). *Adv. Mater.*, **24**, p. 5045.
4. (a) J. R. Miller, P. Simon (2008). *Science*, **321**, p. 651; (b) P. Simon, Y. Gogotsi, *Nat. Mater.*, **7**, p. 845.
5. Z. N. Yu, L. Tetard, L. Zhai, J. Thomas (2015). *Energy Environ. Sci.*, **8**, p. 702.
6. (a) E. Frackowiak (2007). *Phys. Chem. Chem. Phys.*, **9**, p. 1774; (b) H. Zhang, G. P. Cao, Y. S. Yang, B. Xu, W. F. Zhang (2008). *Prog. Chem.*, **20**, p. 1495; (c) Y. P. Zhai, Y. Q. Dou, D. Y. Zhao, P. F. Fulvio, R. T. Mayes, S. Dai (2011). *Adv. Mater.*, **23**, p. 4828.
7. G. A. Snook, P. Kao, A. S. Best (2011). *J. Power Sources*, **196**, p. 1.
8. Y. Wang, J. Guo, T. F. Wang, J. F. Shao, D. Wang, Y. W. Yang (2015). *Nanomaterials-Basel*, **5**, p. 1667.
9. M. S. Balogun, W. T. Qiu, W. Wang, P. P. Fang, X. H. Lu, Y. X. Tong (2015). *J. Mater. Chem. A*, **3**, p. 1364.
10. M. Chhowalla, H. S. Shin, G. Eda, L. J. Li, K. P. Loh, H. Zhang (2013). *Nat. Chem.*, **5**, p. 263.
11. (a) A. Kumar, R. K. Singh, H. K. Singh, P. Srivastava, R. Singh (2014). *J. Power Sources*, **246**, p. 800; (b) Q. Wang, J. Yan, Z. J. Fan, T. Wei, M. L. Zhang, X. Y. Jing (2014). *J. Power Sources*, **247**, p. 197.
12. S. K. Balasingam, M. Lee, B. H. Kim, J. S. Lee, Y. Jun (2017). *Dalton Trans.*, doi: 10.1039/C6DT04466B.
13. A. P. Liu, X. Y. Chen, Z. J. Zhang, Y. Jiang, C. W. Shi (2006). *Solid State Commun.*, **138**, p. 538.
14. Y. F. Shi, C. X. Hua, B. Li, X. P. Fang, C. H. Yao, Y. C. Zhang, Y. S. Hu, Z. X. Wang, L. Q. Chen, D. Y. Zhao, G. D. Stucky (2013). *Adv. Funct. Mater.*, **23**, p. 1832.
15. (a) S. K. Balasingam, J. S. Lee, Y. Jun (2015). *Dalton Trans.*, **44**, p. 15491; (b) S. K. Balasingam, J. S. Lee, Y. Jun (2016). *Dalton Trans.*, **45**, p. 9646.
16. N. S. Arul, J. I. Han (2016). *Mater. Lett.*, **181**, p. 345.
17. P. Pazhamalai, K. Krishnamoorthy, S. J. Kim (2016). *Int. J. Hydrogen Energy*, **41**, p. 14830.

18. D. Chakravarty, D. J. Late (2015). *RSC Adv.*, **5**, p. 21700.
19. N. Moloto, M. J. Moloto, M. Kalenga, S. Govindraju, M. Airo (2013). *Opt. Mater.*, **36**, p. 31.
20. S. Selvam, B. Balamuralitharan, S. N. Karthick, A. D. Savariraj, K. V. Hemalatha, S. K. Kim, H. J. Kim (2015). *J. Mater. Chem. A*, **3**, p. 10225.
21. A. Sobhani, M. Salavati-Niasari (2014). *J. Alloys Compd.*, **617**, p. 93.
22. X. D. Liu, J. M. Ma, P. Peng, W. Z. Zheng (2009). *J. Cryst. Growth*, **311**, p. 1359.
23. D. W. Lee, L. De Los Santos, J. W. Seo, L. L. Felix, A. Bustamante, J. M. Cole, C. H. W. Barnes (2010). *J. Phys. Chem. B*, **114**, p. 5723.
24. H. Wang, H. B. Feng, J. H. Li (2014). *Small*, **10**, p. 2165.
25. J. H. Guo, Y. T. Shi, X. G. Bai, X. C. Wang, T. L. Ma (2015). *J. Mater. Chem. A*, **3**, p. 24397.
26. (a) A. Ramadoss, S. J. Kim (2013). *Carbon*, **63**, p. 434; (b) M. Lee, S. K. Balasingam, H. Y. Jeong, W. G. Hong, H. B. R. Lee, B. H. Kim, Y. Jun (2015). *Sci. Rep.-UK*, **5**.
27. A. Ramadoss, B. Saravanakumar, S. W. Lee, Y. S. Kim, S. J. Kim, Z. L. Wang (2015). *ACS Nano*, **9**, p. 4337.
28. S. K. Balasingam, A. Thirumurugan, J. S. Lee, Y. Jun (2016). *Nanoscale*, **8**, p. 11787.
29. M. F. Cabral, M. L. Calegaro, S. A. S. Machado (2012). *RSC Adv.*, **2**, p. 2498.
30. A. Gobeaut, L. Laffont, J. M. Tarascon, L. Parissi, O. Kerrec (2009). *Thin Solid Films*, **517**, p. 4436.
31. Z. A. Zhang, Y. Fu, X. Yang, Y. H. Qu, Z. Y. Zhang (2015). *ChemNanoMat*, **1**, p. 409.
32. G. D. Park, Y. C. Kang (2016). *Chem. Eur. J.*, **22**, p. 4140.
33. M. Jana, S. Saha, P. Khanra, P. Samanta, H. Koo, N. Chandra Murmu, T. Kuila (2015). *J. Mater. Chem. A*, **3**, p. 7323.
34. B. Muller, H. D. Lutz (1991). *Solid State Commun.*, **78**, p. 469.
35. (a) I. K. Moon, J. Lee, R. S. Ruoff, H. Lee (2010). *Nat. Commun.*, **1**; (b) C. Lu, P. J. J. Huang, B. W. Liu, Y. B. Ying, J. W. Liu (2016). *Langmuir*, **32**, p. 10776.
36. Z. J. Jia, B. G. Wang, Y. Wang, T. Qi, Y. H. Liu, Q. Wang (2016). *RSC Adv.*, **6**, p. 49497.
37. S. Y. Liu, Y. G. Zhu, J. Xie, Y. Huo, H. Y. Yang, T. J. Zhu, G. S. Cao, X. B. Zhao, S. C. Zhang (2014). *Adv. Energy Mater.*, **4**, p. 1301960.
38. J. E. B. Katari, V. L. Colvin, A. P. Alivisatos (1994). *J. Phys. Chem.-US*, **98**, p. 4109.

39. W. Xiao, W. J. Zhou, T. Feng, Y. H. Zhang, H. D. Liu, L. L. Tian (2016). *Materials*, **9**.
40. A. Ramadoss, T. Kim, G. S. Kim, S. J. Kim (2014). *New J. Chem.*, **38**, p. 2379.
41. (a) L. J. Cao, S. B. Yang, W. Gao, Z. Liu, Y. J. Gong, L. L. Ma, G. Shi, S. D. Lei, Y. H. Zhang, S. T. Zhang, R. Vajtai, P. M. Ajayan (2013). *Small*, **9**, p. 2905; (b) A. Winchester, S. Ghosh, S. M. Feng, A. L. Elias, T. Mallouk, M. Terrones, S. Talapatra (2014). *ACS Appl. Mater. Interfaces*, **6**, p. 2125.
42. S. Selvam, B. Balamuralitharan, S. Jegatheeswaran, M.-Y. Kim, S. N. Karthick, J. Anandha Raj, P. Boomi, M. Sundrarajan, K. Prabakar, H.-J. Kim (2017). *J. Mater. Chem. A*, **5**, p. 1380.
43. A. Ramadoss, B. Saravanakumar, S. J. Kim (2015). *Energy Technol.-Ger.*, **3**, p. 913.
44. N. Kurra, C. Xia, M. N. Hedhili, H. N. Alshareef (2015). *Chem. Commun.*, **51**, p. 10494.
45. (a) Q. Cheng, J. Tang, J. Ma, H. Zhang, N. Shinya, L. C. Qin (2011). *Phys. Chem. Chem. Phys.*, **13**, p. 17615; (b) Q. Cheng, J. Tang, J. Ma, H. Zhang, N. Shinya, L. C. Qin (2011). *Carbon*, **49**, p. 2917; (c) M. A. Bissett, I. A. Kinloch, R. A. W. Dryfe (2015). *ACS Appl. Mater. Interfaces*, **7**, p. 17388; (d) N. Savjani, E. A. Lewis, M. A. Bissett, J. R. Brent, R. A. W. Dryfe, S. J. Haigh, P. O'Brien (2016). *Chem. Mater.*, **28**, p. 657; (e) A. Ramadoss, K. Y. Yoon, M. J. Kwak, S. I. Kim, S. T. Ryu, J. H. Jang (2017). *J. Power Sources*, **337**, p. 159.
46. S. Selvam, B. Balamuralitharan, S. N. Karthick, K. V. Hemalatha, K. Prabakar, H. J. Kim (2016). *Anal. Methods-UK*, **8**, p. 7937.

Chapter 7

Solar Panel Temperature Control System Using IoT

Solar photovoltaic systems are renewable energy sources that are used widely around the world. A drawback is that the efficiency decreases as the temperature of the solar panels increases. To prevent this phenomenon, a cooling fan can be installed on the backside of the solar panel to increase the efficiency. Solar system efficiency also decreases due to weather conditions and unexpected situations. To overcome this problem, an Internet of Things (IoT) system was used to monitor the state of the solar system and control the cooling fan. The core microprocessor used in IoT systems was Arduino. Using Arduino, an IoT system can be implemented simply and inexpensively. The entire system was designed and tested and the efficiency increased by approximately 4.7%. Although it is a small 30 W capacity photovoltaic system, its efficiency is expected to increase on application to a photovoltaic system of more than 1 kW in the near future.

Solar Power and Energy Storage Systems
Hee-Je Kim

ISBN 978-981-4800-13-6 (Hardcover), 978-0-429-45877-4 (eBook)
www.panstanford.com

7.1 Introduction

The increasing population worldwide has led to an increased demand for energy. Fossil fuels are not an ideal source for future energy, and the excessive use of fossil fuels causes environmental problems. Clean and renewable energy sources, such as solar energy, wind energy, geothermal energy, hydrogen energy, and fuel cells, are needed to overcome these issues.[1,2] Despite the sustainability and low cost, renewable energy sources are less reliable and the generation of large quantities of electricity is difficult.[3] These shortcomings need to be overcome and energy production needs to be increased to solve any future energy crisis.[4,5] Among these renewable resources, solar energy is cheaper and more sustainable for practical applications, such as heating homes, lighting, and automobiles. In addition, it can be used at any time through energy storage on being combined with an energy storing system.[6] On the other hand, it has several disadvantages; the method of solar power generation is expensive and weather dependent and solar power is accessible only in daytime. Solar energy is accessible in two ways: electricity and heat. Solar panels use the photovoltaic (PV) effect to convert sunlight into electricity, while solar thermal collectors are used to convert sunlight into heat for water heating.

This study examined solar power generation using solar panels and its challenges during energy production.[7,8] Typically, silicon semiconductors are used in solar panels because of silicon's ideal bandgap of 1.4 eV, with a high absorption coefficient and thermal stability in an open atmosphere.

An unwanted side effect of the encapsulation of solar cells into a PV module is that encapsulation alters the heat flow into and out of the PV module, thereby increasing the operating temperature of the PV module. These increases in temperature have a major impact on the PV module by reducing its voltage, thereby lowering the output power. Recent studies have shown that the efficiency of an amorphous silicon cell decreases by 0.05% for every 1°C increase in temperature and in the case of a crystalline silicon solar cell, the decrease ranges from 0.4% to 0.5%.[9] As shown in Fig. 7.1, the output power decreases with increasing temperature.

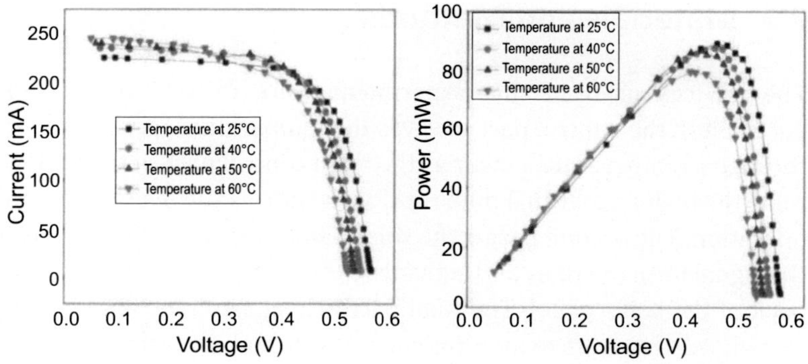

Figure 7.1 Solar panel output power with temperature change.

Although temperature differs according to season and location, the temperature of a solar panel is usually the highest in summer.[10] Various solutions have been proposed to prevent the efficiency decrease from such a temperature rise. First, a cooling fan is used to lower the temperature of the solar panel. The cooling fan cools the solar panel by blowing air onto the back of the panel. Second, the back of the panel can be attached to a heat sink. Third, cooling water is sprayed on the surface of the solar panel, but this requires additional mechanical devices. Among these three methods, the method of lowering temperature using cooling fans was found to be most effective.[12,13] The cooling fan is turned on and off automatically by the temperature of the solar panel. On the other hand, operating the cooling fan in the event of clouds or inverter failure is inefficient. Therefore, in some situations, real-time remote control is needed.

To implement such a system, an operating system based on the IoT is required. The IoT is an ecosystem of connected physical objects that are accessible through the Internet. This is next-generation technology that will soon be applied to society.[14] The IoT refers to the technology or environment in which sensors are attached to transmit data to the Internet in real time.[15,16] On the basis of IoT, the temperature of the solar panel can be controlled anytime and anywhere through a smartphone. In this chapter, the ultimate goal is to automatically control the temperature of the solar panel using an IoT system.

7.2 Materials and Methods

The devices used in this experiment were divided into three parts. First, the control part receives the temperature measurement through a temperature sensor as the main control part, activates the relay through an external interrupt, and controls the overall system operation. The second part is the solar panel side. This part receives the signal from the relay and activates the cooling fan installed on the back of the solar panel. The final part is a smartphone application that allows the users to monitor solar power and control the cooling fan whenever and wherever. The parts and mechanisms of each of these three parts have been described.

7.2.1 Control Part

This part receives the various data values as the main control part and transmits the data to the server in connection with the Internet network. This also activates the cooling fan on the back of the solar panel through an external interrupt. The microcontroller used was Arduino Mega2560 (see Appendix A for the Arduino IED script). This device was optimized to implement the IoT and is economical and can be interworked with a range of sensors. A Wi-Fi shield (JSN270) for implementing the IoT was also used. Tables 7.1 and 7.2 list the specifications of Arduino and the Wi-Fi shield (JSN270).

The power to operate Arduino was obtained from an internal power source produced from a solar panel. The power consumed by the control part was 1.12 W. Figure 7.2 shows the internal view of

Table 7.1 Specifications of Arduino Mega 2560, one of the microcontrollers

Type	Value
Voltage	5 VDC
Digital I/O pins	54 Pins
Analog I/O pins	16 Pins
Flash memory	256 KB
SRAM	4 KB
Clock speed	16 MHz

Table 7.2 Specifications of the Wi-Fi shield (JSN270) for IoT implementation

Type	Value
Wireless standard	IEEE 802.11 b/g/n
Frequency range	2.412 $\approx$ 2.477 GHz
Output power	17 dBm at 1 Mbps, 15 dBm at 11n MCS0, 13 dBm at 11n MCS7
Receive sensitivity	−97 dBm at 1 Mbps, −93 dBm at 11n MCS0, −75 dBm at 11n MCS7
Data rates	Max. 72 Mbps
Modulation type	OFDM, DSSS

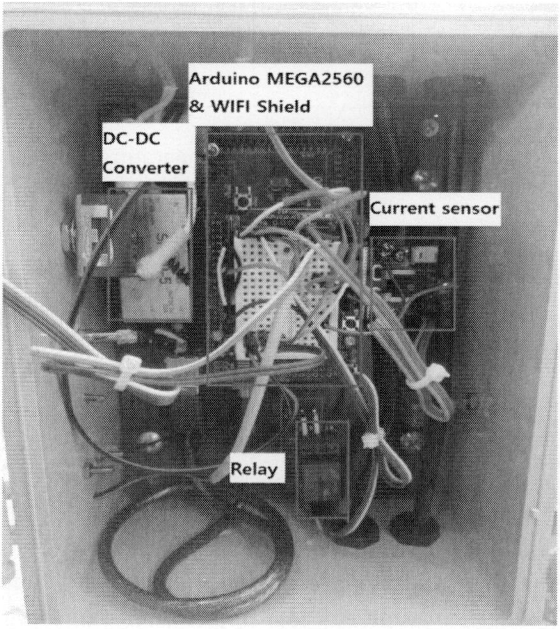

Figure 7.2 Inside the control box with Arduino and a Wi-Fi shield.

the control box. In the center, there are Arduino and the Wi-Fi shield, as well as converters, current sensors, and relays.

Arduino controls the system via the flow diagram given in Fig. 7.3. The cooling fan turns on automatically when the

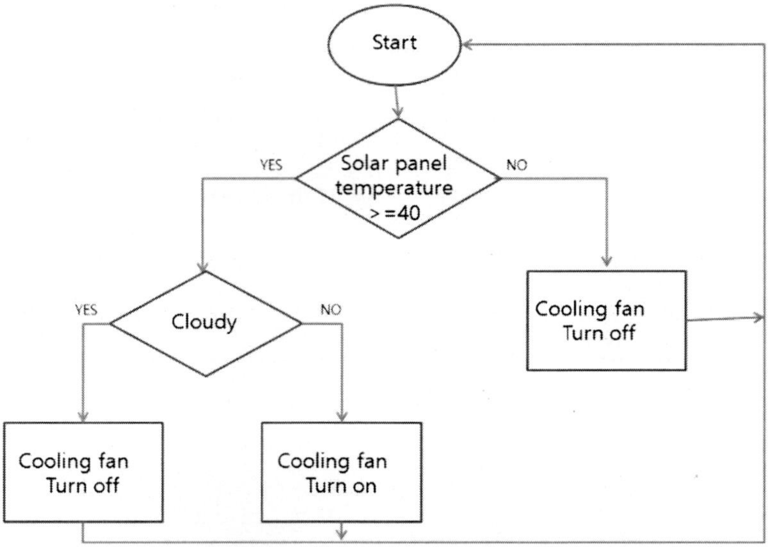

Figure 7.3 Algorithm of the solar panel temperature control system.

temperature of the solar panel exceeds 40°C and when there are no clouds. Figure 7.3 shows the algorithm of the solar temperature control system.

7.2.2 Solar Panel Part

This part consists of a cooling fan and a solar panel. Four cooling fans are installed on the back of the solar panel (Fig. 7.4). This is arranged in a regular square arrangement for efficient cooling. Four cooling fans consume 360 mW of power. Arduino controls the cooling fan through the relay on/off operation. The solar panels had a 30 W capacity for testing and were fixed at an optimal angle of 35°C to maximize the solar energy. In addition, an infrared (IR) temperature sensor was attached to the back of the solar panel to monitor the temperature of the solar panel in real time. Tables 7.3 and 7.4 list the specifications of the cooling fan and the solar panel.

Table 7.3 Specifications of the cooling fan

Type	Value
Fan size	80 mm × 80 mm × 15 mm
Voltage	12 VDC
Input power	90 mW
Speed	3600 rpm
Weight	120 g

Table 7.4 Specifications of the solar panel

Type	Value
Max. power	30 W
Max. power voltage	19.6 V
Max. power current	1.64 A
Open-circuit voltage	24.0 V
Short circuit current	1.77A
Test condition	AM 1.5 [25°C 1000 W/m^2]

7.2.3 Smartphone Application Part

This system does not allow people to go directly to the solar panel to turn the cooling fan on or off. The cooling fan should be controllable at anytime and from anywhere. To implement this, control using a smartphone application is essential. Open source software called App Inventor 2 was used to develop apps. This is not only easy to make but also available for free. An essential element to this application is the part that can monitor the sensor values received from Arduino and the button part to control the cooling fan. Figure 7.5 presents the designed application and the source used in App Inventor 2.

7.3 Experimental Setup

Each component was designed and assembled to produce a model for the experiment. The experiment was carried out from 09:00 to 18:00 in clear, cloudless weather. The location was Pusan National

Figure 7.4 The back of the solar panel with four cooling fans.

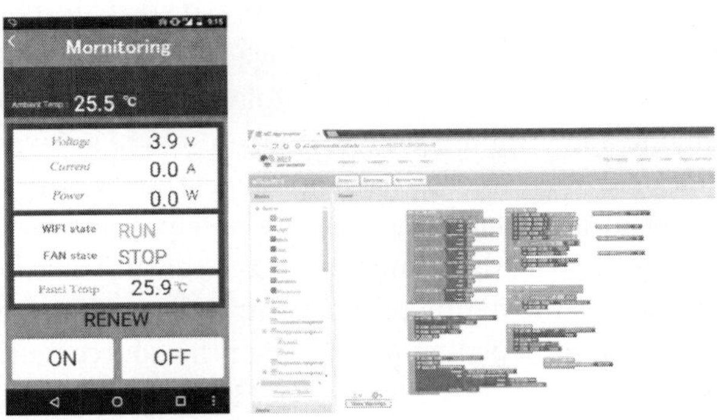

Figure 7.5 (a) Designed application and (b) App Inventor 2.

University, Busan, Korea (31.040°N, 31.486°E), building number 207 roof, and the experiment date was around September 2017. The angle of the solar panel was maintained at 35°C, and the direction of gaze pointed to the south. An IR thermometer was used to measure

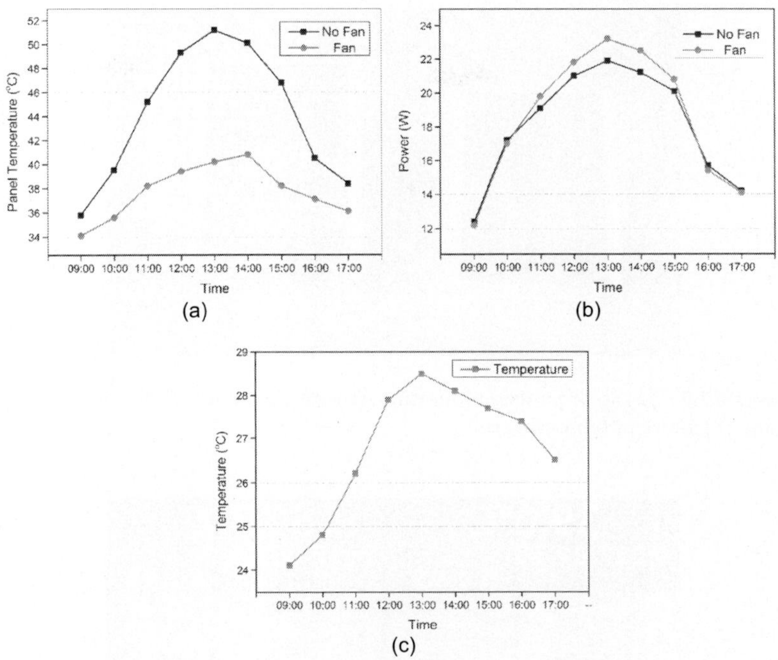

Figure 7.6 Arranged solar panels (Model A on the left; Model B on the right).

the temperature of the solar panel. A VT04A visual IR thermometer (FLUKE, Everett, WA, USA) was used. Model A (without a cooling fan) was compared with model B (with a cooling fan). The system was connected to the network for monitoring and remote control of the cooling fan. Figure 7.6 shows the layout.

7.4 Results and Discussion

The solar panel was fixed in a suitable wide area, and the experiment was then conducted. At the beginning of the experiment (09:00 $\approx$ 10:00) and at the end (16:00 $\approx$ 17:00), the output power was similar because the temperature difference between the two solar panels was small. On the other hand, from 11:00 to 15:00, when the temperature difference of the solar panel was large, the solar panel

Figure 7.7 (a) Solar panel temperature, (b) output power of the solar panel, and (c) ambient temperature.

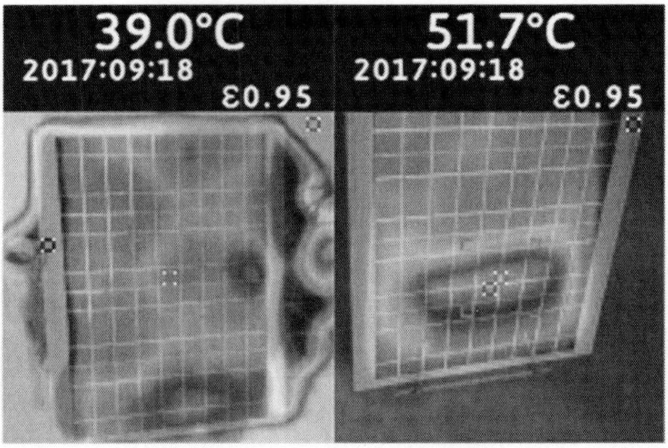

Figure 7.8 Maximum temperature difference of a solar panel.

with the cooling fan showed higher output power than the solar panel without the cooling fan. The difference in the output power reached a maximum at 13:00, when the temperature difference between the solar panels reached its maximum. Figure 7.7a–c shows the solar panel temperature, ambient temperature, and output power over time.

And Fig. 7.8 shows a solar panel image at 13:00, at which the temperature difference between the two solar panels was maximized.

The output power graph shows that the output power can be determined using this system from 11:00 to 15:00. At that time, the increase in efficiency of the output power was approximately 4.7%. This figure can change depending on the season and weather. On the other hand, it is practical to use the IoT system to have control at any time and from any place. Moreover, a higher power may be obtained by applying it to a large-capacity PV system rather than to such a small-capacity PV system.

7.5 Conclusion

The solar PV system becomes less efficient as the temperature of the solar panel increases. In particular, the temperature of the solar panel increases by more than 40° during the daytime (11:00 $\approx$ 15:00). Therefore, a cooling fan was installed on the back of the solar panel to increase the output power. On the other hand, the solar system efficiency can become low because of hazy conditions and other reasons. If the temperature can be checked in real time and the cooling fan can be controlled automatically, the efficiency can be improved again. Therefore, real-time remote control through an IoT system was implemented to increase solar power efficiency. The IoT system device used a microprocessor called Arduino, which makes it possible to construct an inexpensive and simple system. This system is compatible with various sensors and can implement various functions. Finally, after designing the IoT system and solar system, a successful experiment was conducted. Using a 30 W solar panel, the maximum increase in efficiency was approximately 4.7% and the power increased by 1.4 W. No decrease in efficiency was observed because the system operates with the power generated from the solar panels. If the system is applied to a large-capacity PV system, higher power can be effected by the cooling fans. Future studies will extend the system and facilitate its application in a 1 kW PV system.

References

1. K. Branker, M. Pathak, J. Pearce (2011). A review of solar photovoltaic levelized cost of electricity, *Renewable Sustainable Energy Rev.*, **15**, pp. 4470–4482.
2. M. Child, C. Breyer (2016). Vision and initial feasibility analysis of a recarbonised Finnish energy system for 2050, *Renewable Sustainable Energy Rev.*, **66**, pp. 517–536.
3. D. P. Schlachtberger, S. Becker, S. Schramm, M. Greiner (2016). Backup flexibility classes in emerging large-scale renewable electricity systems, *Energy Convers. Manage.*, **125**, pp. 336–346.
4. L. Fontanet, J. García (1996). Gestión del medioambiente urbano. Residuos que se generan en la actividad de la construcción. Directiva U.E. "Demotion Waste", *Inf. Constr.*, **47**, pp. 137–144.
5. J. M. Pearce (2002). Photovoltaics: a path to sustainable futures, *Futures*, **34**, pp. 663–674.
6. P. Sathishkumar (2017). A blended SPS-ESPS control DAB-IBDC converter for a standalone solar power system, *Energies*, **10**(9), p. 1431.
7. B. Kim, K. Kim, K. Kim (2017). Determining the optimal installation timing of building integrated photovoltaic systems, *J. Cleaner Prod.*, **140**, pp. 1322–1329.
8. H. Batoul (2008). Flow simulation improves photovoltaic solar panel performance, *Technical Report*, Schueco International, Paris, France.
9. S. Dubey, G. N. Tiwari (2008). Thermal modeling of a combined system of photovoltaic thermal (PV/T) solar water heater, *Sol. Energy*, **82**, pp. 602–612.
10. Y. A. Cengel, M. A. Boles (2005). *Thermodynamics: An Engineering Approach*, 5th ed., McGraw-Hill Science, New York, NY, USA.
11. J. K. Tonui, Y. Tripanagnostopoulos (2007). Improved PV/T solar collectors with heat extraction by forced or natural air circulation, *Renewable Energy*, **32**, pp. 623–637.
12. M. A. Khan, B. Ko, E. Alois Nyari, S. E. Park, H.-J. Kim (2017). Performance evaluation of photovoltaic solar system with different cooling methods and a bi-reflector PV system (BRPVS): an experimental study and comparative analysis, *Energies*, **10**, p. 826.
13. B. Zinber, G. Makrides, W. Schmitt, G. E. Georghiou, J. H. Werner (2007). Annual energy yield of 13 photovoltaic technologies in Germany and in Cyprus, in *Proceedings of the 22nd European Photovoltic Solar Energy Conference*, Munich, Germany.

14. AMECIT. Informe de smart cities 2012. Available online: http://ametic.es/es/publicaciones/smart-tics (accessed on 5 July 2017).
15. C. Doukas (2012). *Building Internet of Things with the Arduino*, 1st ed., CreateSpace, North Charleston, SC, USA, pp. 1–352.
16. M. Kuniavsky (2010). *Smart Things: Ubiquitous Computing User Experience Design*, 1st ed., M. James, ed., Elsevier, San Francisco, CA, USA, pp. 1–287.

Index